I0030121

ONE SECOND MATH

ONE SECOND MATH

HOMEWORK HELP ON DEMAND

MATHCELEBRITY

COPYRIGHT © 2016 MathCelebrity

All rights reserved.

ONE SECOND MATH

Homework Help on Demand

ISBN   978-1-61961-479-6   *Paperback*

       978-1-61961-480-2   *Ebook*

LIONCREST
PUBLISHING

I want to thank my wife Angela for backing my play on this journey. Her support on a project of this size is commendable. This website gave me a 1,000 mile stare thinking about mathematical pattern recognition. I thank her for her patience with my endless hours of computer time and being a voice of reason at the end of the day. My wife has helped our family as a representative selling Poofy Organics products (https://angelasevcik.poofyorganics.com).

To my daughter Stella who arrived after I finished a majority of the code. I want to make a better life for you and spend more time with you.

To my Father-In-Law and Mother-In-Law, Frank and Shirley McLaughlin. I consider you my parents. Your support throughout this mental marathon has been incredible. The true mark of accomplishment is when others give you unsolicited praise. There is no better example of this then both of you.

To my Brother-In-Law and Godfather to my daughter, Daniel McLaughlin. Daniel is the CPO of MathCelebrity.com and a Young Living Independent Distributor (Member ID 1933752). Here is to late night calls going wild over traffic numbers. Here is to dreaming about leaving the Grind someday.

To my good friend Rich Mavec, one of the few people I can have an intelligent conversation with. Entrepreneurship is a lonely road, and I enjoy our talks on business and politics.

To my lawyer and friend, Bernard G. Peter, for being a voice of calm reason and encouragement. Bernard is one of the few people on Earth I trust to discuss just about anything.

To Barry Codell, one of the brightest minds I have ever been fortunate enough to meet. Working on the BarryCode.com website with the guy who inspired the MoneyBall book has been a pleasure.

To my friend David Kimrey, one of the few people I always learn great business and marketing ideas from. David is the cofounder of MavenX.com a site that is like Pinterest but pays you when people purchase from your posts.

Thanks to the GodMother of my daughter and my wife's best friend, Tina Wisnoski. We have had great discussions on marketing and ideas for exam certifications.

Thanks to Sambazon Organic Energy Drinks for morning brain power.

Thanks to Whole Foods and Chipotle for being my meals of choice.

* * * * * * * * * * * * * * * * * * *

CONTENTS

CHAPTER 1

INTRODUCTION

This book is the culmination of eight years and five hundred thousand–plus lines of coding.

The following book covers eight years of work to build the MathCelebrity website. Topics covered include programming, website building, sales, marketing, and corporate politics. This book is split into two parts—the website story followed by a large book of math formulas and shortcuts. The website story is structured in three parts—what, why, and how.

1. What motivated me to build MathCelebrity?
2. Why did I build MathCelebrity?
3. How did I build MathCelebrity?

The title "One-Second Math" is the culmination of my ultimate goal—building an automated math tutor to solve math problems in one second or less.

One Second Math comes from an article I read about the human eye. It takes one-third of one second to blink. Blinking is both rapid and automatic. Rapid and automatic is exactly how I built MathCelebrity.

I rounded up the blinking time to one second, and voila, I found my book title.

ORIGINS

I'm sitting at a mind-numbing job in a dreary cube farm. With five hours to go before closing time, I dream about a better outcome for the workday. In ten minutes, I will attend another useless meeting. Ah, the meeting—I challenge you to find a bigger time waste in corporate America. If no agenda is set, you best believe it will be a gab fest where little to nothing gets accomplished.

This particular meeting began, and I started daydreaming immediately. I daydreamed frequently at my day job. At this point in my career, I earned a decent salary at thirty-one years old. I recently jumped ship from a competitor. I missed the friends I worked with, but I had to leave. Over the last few years, I learned a valuable lesson never taught by career counselors: you make more money jumping ship to another job than you make sitting around waiting for a raise. I call this sitting-around-and-waiting game "chasing the carrot." Remember the term *chasing the carrot*; I cover it later in the book.

A bit of background: I am a programmer by trade. For the past twenty years, I have worked in employee benefits and actuarial science. I specialize in automating anything and everything possible. I love writing a program where you push a button and something immediately happens. Even better than pushing a button? An automated, scheduled task you don't have to think about. You sit back, have breakfast, and blow people away. Zero stress, zero worries, instant results.

My ruthless automation approach created friction in corporate environments. In many corporations, everything is "by the book." If you never worked in a corporate job, allow me to paint this picture for you. Imagine a world where you need five levels of approval and a form filled out before you go to the bathroom. Boring, dreary colors are rampant. Innovation is stifled and ingenuity gets squashed. To top it all off, the raise and bonus process is laughable. Yet in the corporate world, you can't fight it. It's a constant drag pulling you down, just like gravity.

Because I don't fight gravity, I needed to branch out. I wanted to earn more money outside the corporate spider web. To make more money, I started hustling with side gigs in programming and math tutoring. Let me pause for context: Take a $60,000 salary at a day job. If you get a 3 percent raise, you earn $1,800. 3 percent is equivalent to an average cost of living increase. To get this raise, you must work all year without any blemishes on your work history.

Let's compare your day job to a side business. You can make $1,800 quite easily with a few hours per week of side hustle and solid networking. Assume you find something to work on at $30 per hour. You find five free hours each week before or after work. You earn $150 extra per week. If you work on this side gig for 3 months, you make $1,800. Even better, you avoid annual reviews and drama at work. This is a reasonable return on your time investment compared to your day job.

I pursue anything with a reasonable return on investment (ROI) and limited time investment. I hate wasted time. A side hustle with a guaranteed ROI requiring a few hours a week is attractive to me. I prefer things I can control.

With ROI in mind, I started my side hustle. I took on math tutoring as my first gig. Because I majored in math in college, this became the obvious choice. I signed up with a tutoring marketplace. One client turned into two. Two clients turned into four. The math tutoring business produced lucrative returns and happy clients. My clients' referrals grew my side business.

I met clients through mutual friends and online tutoring hangouts. As my tutoring business flourished, it consumed more of my time. I had to turn down a few appointments to avoid conflict with my day job. Flushing these opportunities down the toilet killed me inside.

As time went on, my day job got busier. I found more opportunities online to grow the side business, but scaling up presented a challenge. I'd have to put in twelve or more hours each day, leaving no time to spend with my wife. I had to do something. I refused to let this lucrative side hustle die a slow death.

My side gig gave me the freedom to make money at will. I did things my way on my schedule. One problem remained—I wanted to reach more students and be available more hours. I love automation, I enjoy tutoring, and I want to expand. I want to build something meaningful. I needed a solution to this problem—how do I expand? In a twisted sense of irony, I found the answer to my problem at my day job.

CHAPTER 3

LIGHTNING STRIKE IN INDIA

The alarm blared the next morning signifying another boring day at work. My boss asked me to build a spreadsheet for a colleague in India. This spreadsheet needed crystal-clear instructions for our colleagues across the pond. We simplified the spreadsheet into three steps:

- Enter details.
- Push a button.
- Thousands of lines of consolidated reporting get e-mailed within thirty seconds.

This spreadsheet contained hundreds of summary reporting calculations. My colleagues in India wanted to learn. However, they received basic training documents. They needed more detail. They wanted to learn on the fly.

I made sure the numbers and descriptions were easy to follow on this spreadsheet. I got off to a good start, but the spreadsheet needed more. As numbers and parameters change, so do the details. Therefore,

I decided any cell with a formula needed another cell with supporting math. As the spreadsheet changed, the math work updated immediately. My colleagues in India now had a dynamic learning tool. Each cell value had supporting detail.

My coworkers in Bangalore were thrilled. I built a simple, quick, and easy-to-follow program for them. The Indian team asked me to build more spreadsheets exactly like this. The next spreadsheet I built had multistep formulas needed to get to an answer. I placed math formulas in parentheses in one cell next to the answer cell. Updates were instant like this:

Cell Formula and Math Work	Cell Result
Pension Amount:	$250.00
Offset Percentage:	30%
Reduction Amount ($250 × 0.30):	$75
Reduced Value ($250.00 – $75.00):	$150

When Pension Amount or Offset Percentage changes, the Reduction Amount and Reduced Value cells update. Also, the math work updates automatically. My colleagues loved it. I cut their learning curve in half overnight.

The spreadsheet had one more missing puzzle piece: validation. When values changed, we needed to ensure these changes were valid. I did this by highlighting important cells based on value ranges. Over the next week, I added color coding and formatting changes in cells based on values.

Color	Meaning
Red	Negative values or values out of range
Yellow	Values approaching a dangerous level
Green	Values are correct

I put the finishing touches on a spreadsheet one night and stared at the screen. I thought to myself, *this looks and acts like a math problem.*

The lightbulb in my head turned on. What if I built an automated math tutor? This included the step-by-step math work where anybody understood it. I planned to build it like the spreadsheets at work. You enter your problem, push a button, and the work instantly shows. This served my clients without being face-to-face with them. Because my colleagues in India followed this process, so could my tutoring clients. This became the birth of the MathCelebrity website. I decided to build an automated math tutoring solution like the spreadsheets.

Next day at work, I went to lunch, and the entire time I ate, I kept thinking about the math tutoring website. Ideas popped into my head. I returned from lunch and Googled math tutoring websites. I found a few websites with online notes and simple calculators. I came across zero websites displaying the step-by-step work with an easy-to-use menu. Nothing resembled what I built in my Excel spreadsheets. Other educational websites I found had a confusing layout. Users had trouble finding what they needed. This presented an opportunity for me. I skimmed through a few online forums to see what people said about online math tutoring. Three questions popped in my head from this research:

- What subjects will the website cover?
- How will the design look? One calculator per concept?
- How will I build the search engine and navigation?

I avoided building a big mind map. Instead, I wanted to find answers to my three questions. These answers formed the basis for the initial website setup. My personality gears toward building something first and evaluating results on the fly. Remember the entrepreneurial saying: ready, fire, aim? Entrepreneurs cross the street while average Joes wait for stoplights.

I hate analysis paralysis. Sitting around pondering is a giant waste of time to me. To get the process moving, I set the first goal for the website: build one simple addition and subtraction calculator. Next, get visitors on the website, see how they liked it, and go from there.

First things first, I needed a name. I found an idea reading through the tabloids. I remember seeing reports of celebrities doing outrageous stuff in public. In spite of this, the general populace worshiped their affairs. It seemed trendy to bash people who enjoyed learning. Concurrent to this mindset, another depressing statistic appeared. American math scores declined over the previous years. Put these scores on a graph and observe the downward slope getting worse by the minute. A once-proud country who led the world in math now fell behind at an embarrassing rate. Oh, how far America had fallen!

Then it dawned on me: we can make intelligence cool again. I envisioned a red carpet where visitors get their math homework served up on a platter. It made them feel like a celebrity at a big event. Celebrities get things handed to them every day. I wanted to treat math homework the same way. Press the button, get immediate help. No questions asked.

MathCelebrity—the name stuck. I told my wife and she agreed. Our initial motto declared, "Rolling out the red carpet for you in math." We decided to abandon this. Yet the original premise of making math fun, easy, and served up on a platter stuck.

Shortly after choosing the company name, I came up with our motto. The motto incorporates the four basic math operations. I replaced the numbers with three nouns and one verb. This motivational math operation resulted in math success.

Here is the equation describing our mission: Add Positivity, Subtract Doubt, Multiply Knowledge, Divide Time Spent. Add positivity, because confidence conquers challenges. Subtract doubt, because belief brings relief. Multiply knowledge, because learning begets more learning. Divide time spent, because the website improves student learning speed.

With our motto set, I purchased the domain www.mathcelebrity.com in July 2007. Our name, story, and destiny began.

Yet the eight hundred–pound gorilla in the room remained: I had no clue how to build websites.

BUILDING BRICKS

In order to build my dream, I had to learn a web programming language. My programming background consisted of Excel spreadsheets and Visual Basic. I'd never built a website before; at this point, it was a dream. Dreams are great, but execution reigns supreme. I spent the next three days using every moment of my free time learning how to set up a basic web page. Page layouts, text, headers, images, you name it—the entire process captivated me.

I started with a basic HTML web page. I built two input boxes: one for each number. I placed a button beneath the boxes called Add. I placed a digital notebook paper directly beneath the Add button. The notebook paper contained the math work supporting the answer. One problem existed: event capture. HTML does not allow you to capture events such as button pushes. When a button is pushed, I needed the step-by-step math work to populate.

I researched more and found a language called JavaScript. JavaScript captured events such as the push of a button. With a little practice, I found what I needed.

After learning Javascript basics, I began building the first calculator, adding two numbers. You enter your numbers, push the button, and the step-by-step math work appears. Four hours later, it worked. I built a lightning-fast, yet ugly, page. I have never been a design guy. It's a running joke between friends and family. I build Ferrari engines with a Yugo paint job. I prefer function over form.

One week passed after building the first calculator. I checked the web statistics to see if anybody stumbled across the website. The statistics showed eleven visitors. I cheered because anything over zero visitors indicated progress. At least somebody dropped by.

Over the next two weeks, I built more calculators. Things became easier as I built out the site. I focused on calculators with simple math. This helped me learn how to program from the bottom up. I read through a basic math book in my basement and built everything possible: multiplication, division, numerical properties, midpoints, tallies, and the list goes on. Building the easier calculators first prepared me for difficult calculators later.

As I created more-difficult calculators, my programming skills improved. This helped me tackle more complex challenges. After I completed basic math, I moved on to more advanced subjects: pre-algebra, geometry, and trigonometry. After two months, the website had one hundred calculators in these three subjects. I pressed forward finding more math problems to build.

One day while hanging out in the laundry room, I noticed a few boxes under a table. When I rummaged through these boxes, I found old math books from college. The box contained the following books:

- Statistics
- Linear algebra
- Calculus

Over the next six months, I built more calculators in the morning, at night, and on the weekends. It transformed into an obsession. As much work as I completed, it never felt like work. For my day job, just getting out of bed every day became a struggle. I had zero motivation, engagement, or reason to care. Get up, shower, eat breakfast, go to work, attend silly meetings, rinse, and repeat—this was the grind I dreaded.

My wife commented on my sighing with despair as the alarm went off each work morning. It became a constant struggle dragging myself into another day of the grind. If you have ever seen the old Dunkin' Donuts commercials, you know what I'm talking about. "Time to make the donuts!" That commercial sums up the daily-grind perfectly. Yet waking up on a whim at 3:00 a.m. to work on my website seemed easy and felt right. I pressed forward building the website. I treated my day job like an inconvenience—finding any excuse available to work on my website.

I worked on MathCelebrity quietly. Nobody knew about this project except my wife. At this point, I treated my website like a hobby. A dedicated, time-consuming, unapologetic hobby. It got to the point where my wife and dogs walked downstairs to the office if they wanted to find me. Take the stairs into the basement, go straight back to the wall, and turn right into the office. They found me huddled behind a computer in the back corner. This website consumed a majority of my free time. I wanted to keep moving ahead to build our audience.

I checked the web statistics, and I saw our traffic continuing to climb. Nothing massive, but a positive gain nonetheless. Visitors spent more time on the website, they shared more links, and our rank moved up. As the momentum built, the entire project became a positive feedback mechanism. I built a new calculator, celebrated, and moved on to the next one. I finished an entire math book, drank a Rusty Nail, and kept pushing forward. Every time I built a new calculator, traffic increased and more people talked about us. I kept building more features, we kept getting more traffic. The feedback loop continued.

During this time, I worked with blinders on. I ignored web design and promotion. I ignored suggestions and only built calculators. When you build a business, everybody has an opinion. It pays to be stubborn sometimes. Being stubborn is one of my personality traits. Sometimes it hurts, and sometimes it works to my advantage. For this project, it became a virtue. I'm reminded of the scene in the movie *Deep Cover*. Laurence Fishburne plays an undercover cop posing as a criminal. His boss tells him, "In this job, all your faults become virtues." Impatience and stubbornness turned into an asset on this project.

I pushed forward with one goal in mind: build as many calculators as imaginable, as fast as possible. After nine months, I had 150 calculators under my belt. I learned a pile of HTML and PHP programming in the process. I began this journey with no web programming skills. Now, I had a functioning website.

I hit the brakes for two days on the website to focus on my day job. This resting period allowed me to get feedback from our fans. Around this time, a few friends and my brother-in-law asked me about the project. I explained the website and showed them what I had.

Their verdict: great idea, ugly design. Because I ignored web design, this critique made sense. I shrugged off web design changes since our traffic kept growing. While I ignored web design changes, our fans requested a calculator design change. Next on my list to add was math notation.

One thing I noticed is JavaScript does a poor job displaying math symbols. I wanted a professional textbook layout for the math work. Math notation is essential for step-by-step solutions.

Our fans requested another calculator design change – concept searches. You start typing a concept such as "equation," and the drop-down populates best matches. This saved people typing time and eliminated spelling errors. You know how Google's instant search works? I wanted a MathCelebrity instant search.

I scoured the web to research possible solutions. I stumbled across a programming language named PHP. PHP provided two things:

1. Math notation symbols
2. On-Demand search

In order to use PHP, I had to make a difficult decision. Imagine ripping a bandage off your skin. Painful, but necessary. To accomplish my goal, I needed to rewrite the entire website.

I chugged a Sambazon energy drink and contemplated what I needed to rewrite this website—weeks, months, years? I hired a former coworker to teach me the basics. My coworker is a PHP wizard. We went over the action plan, and he taught me some quick tips to make the process go as smooth as possible. He cut my learning curve down by months. The more I read through the code, the more some patterns started to emerge. I discovered an easy translation of variables, loops, and functions from JavaScript to PHP.

I got to work writing this program. How does it work? It reads the JavaScript code and picks off pieces. If the pieces between the languages are the same, it keeps them. If the pieces are different, it translates them to PHP. You know how programs like Google Translate work, translating from one language to another? This is exactly how my translation program worked. It takes JavaScript, and translates it to PHP. Here's an example:

- Variables in JavaScript are written as var number = 2.
- Variables in PHP are written as $number = 2.

- The translation in my software changes "var" preceding any variable to "$" in front of the variable.

At the end, it pieces them back together again with the push of a button. It is code that writes code. Crazy? I know, but it worked.

I spent eight hours programming during a warm Illinois Saturday. I wanted to be outside in the sunshine, but the translation project consumed my time. My adrenaline kept pumping, so I kept programming. I missed a full summer day, yet I had zero regrets. Just before dinnertime, I rewrote the first calculator. I used my translation program and converted it over with the push of a button. After converting three calculators, I found little tweaks to make the process easier.

Once I had this automatic coder humming, I set a personal goal: five calculator translations per day. I am a big believer in *kaizen*—a small, constant improvement per day every day. With 120 calculators remaining, I needed twenty-four days to complete a full rewrite of the website. I planned to roll out the new changes for the new school year.

I looked at the clock—9:30 p.m. Sunday night. I felt a hint of anger because my day job got in the way of progress tomorrow.

I woke up Monday, and I blazed through my silly day job work in two hours. Management expected a progress report by midweek on a priority project. After finishing my work at breakneck speed, I handed it off to my teammates for review.

I planned on spending time translating calculators while my teammates reviewed my work. What was the harm? Well, this small amount of time turned into five hours with no lunch. Five hours with no lunch turned into twenty calculators rewritten using the new language. Nobody at my day job had a clue. As long as deadlines were met, nobody complained.

I picked my head up from my desk and glanced at the clock—4:00 p.m., one hour from closing time. I helped a coworker with another day-job task working on some Excel automation. I spent 15 minutes on this task, and quietly snuck away for 45 minutes to work on MathCelebrity. I left work at 5:00 p.m. elated with my progress.

Next day at work repeated itself—day-job tasks, MathCelebrity calculator rewrites, go home, eat dinner. By the end of the week, I rewrote fifty calculators. Coworkers at my day job were happy. Blending a day job and side business seemed easy. At this rate, working in secret at my day job, I shaved my original time estimate of twenty-four days in half.

Halfway to my goal, I wanted to see how this change affected the website. I checked the web reports and smiled. People shared the calculators more. Fans stayed on the site longer. It appeared the rewrite of the website reaped many benefits. I closed out the work week a happy man.

I woke up at 4:30 a.m. on the weekend, which meant no day job. Oh, what a joy! I got to work on the calculators. I checked the web reports in the morning. The web statistics improved with more traffic and more people sharing links to our site. MathCelebrity started getting over one-hundred visitors per day. This represented a tenfold increase from the prior year. More traffic meant more feedback.

A parent wrote me nice feedback on how the website helped her son in algebra. She said the step-by-step work helped her son move up one letter grade. I was in heaven. When you start a project like this, you never know if people will like it or won't care. Getting e-mails like this propelled me forward. There is no better feeling in the world than doing something you love. This website is all I wanted to do for work. Each improvement I made to the website gave us more visitors and shares. More visitors beget more visitors.

These constant improvements followed by fan appreciation represented a positive feedback loop. If you have not heard of a positive feedback loop, let me describe it for you. Each small positive victory led to more positive victories. As momentum builds, small success will coalesce into big progress.

It's a powerful thing to be a part of once it happens. It is *kaizen* personified, except for one small difference. The 1-percent improvement per day I talked about went out the window. It turned into 20 percent. I still remember plowing through calculator rewrites until 10:00 a.m. on Saturday. I woke up, skipped breakfast and knocked out thirty calculators.

Ambition always amazes me. When you do something you enjoy, you have an unlimited pile of energy you can tap into. When your brain and body fire on all cylinders, you feel unstoppable. Sadly, your energy pool vanishes when it's time to do something you dislike.

Later that night, my wife and I went out with some friends, and came home at 1:30 a.m. I fell asleep right when we got home. While sleeping, I dreamt about the website. Have you ever seen the movie *Limitless*, where Bradley Cooper "sees" the book before he writes it? That's exactly how my dreams were. I saw the code changes in my dreams. I leaped out of bed at 5:00 a.m. and attacked the site once more.

Angela slept in, so I worked on the website another five hours until she woke up. I had twenty more calculators to go until completion of the upgrade. I planned to finish them on Monday.

Just one problem—my heart sank because I had to work the next day.

DAYLIGHTING

During the first five years, I worked on this website all the time. I pored through books, notes, tickets, and ideas to grow this thing. My family worried about me burning out working a full-time job and a side business. When people find out I have a day job and MathCelebrity, they can't believe it. They ask, "How do you juggle a website of this size with a day job?"

As you read last chapter, you may know the answer. Let me set the stage for you first. While I enjoy programming, I dislike corporate politics. My job paid bills, but the lack of meritocracy in the workplace frustrated me.

Call me a dinosaur, but I believe in pay for performance. Small talk, hobnobbing, and jealousy should never find a way to your annual review.

Alas, the corporate world works in strange ways. Office politics exists everywhere; it is a life-force drainer. I learned an important lesson: there are very few meritocracies left in corporate America. I note the following exceptions below:

1. Sales and commission-based positions
2. Stock/bond/options trading

3. Running your own business

After working for seven years, I developed a process in my career. I changed jobs, stayed for two or three years, and then I left. At every job, a repeating pattern emerged. I worked hard for one year, automated everything, optimized software, and smashed deadlines. One year into any job, I stopped and waited for my review. One of two things always happened:

1. My review occurred on time, and I received a small raise and bonus, or no raise and bonus.
2. Management pushed back my review where the delay cost me more money.

If option 1 happened, I did the required work and no more. By required, I mean enough to make people happy without overexerting myself. Because what does doing extra get you? Nothing. Let's say your written review contained strong feedback. Logical thinking dictates high performance sets you up for a raise. However, logic vanishes in corporate America. Instead, your raise is equal to a cost of living increase, sometimes less. If your raise equals the inflation amount, what is the point of working harder?

In my twenties, I got mad. But with age comes wisdom. I learned job jumping gets you far more money than sitting around collecting cost-of-living increases. Instead of 3 or 4 percent, job jumping gets you 20 percent or more. Once I learned this lesson, my average time on a job amounted to two years. When two years hit, if the opportunity to advance dwindled, I left for more money elsewhere. Why hope for an annual review to go your way?

Hopium is a dangerous drug. Instead of hoping, I wanted to learn the truth about reviews. A mentor took me aside when I started

MathCelebrity. He removed the veil surrounding the mysterious annual review process. He taught me how things really work. For years, I assumed strong performance equals a healthy raise. Oh, how wrong could I be! Allow me to share with you the mystery of the annual review.

Buried deep within your company is a chosen group of people. Let us call them the Inner Circle. The Inner Circle consists of managers and decision makers. You may think the Inner Circle reviews your annual feedback and makes a decision, but you are mistaken. Here is where office politics rears its ugly head.

The Inner Circle gets together privately in a secluded room each year. They pick names off a list in your company one by one. The review leader calls out the next name: "Don Sevcik's review is next."

- Person 1 says, "Don saved us a lot of time with the programs he built."
- Person 2 says, "Don's automation helped the company grow."
- Person 3 says, "Don helped out on extra projects."

Person 1, 2, and 3 recommend a raise of 7 – 10%. So far, so good, right? Wrong. Person 4, who never worked with me during those twelve months chimes in: "I heard Don did a great job this year, but I just don't know much about him. I don't see him at many social events after hours. I don't talk to him much, so maybe we give him a cost of living raise."

After everybody in the Inner Circle chimes in, a decision gets made. My raise gets cut from 10 percent to 3 percent because person 4 has significant pull in the company. Poof, gone, my raise vanishes like a thief in the night. Have you ever seen the movie *The Usual Suspects* with Kevin Spacey? Poof, and like that . . . he's gone.

For no performance-related reasons, I take a 50-percent lump to the head on a raise and bonus. Let the meritocracy be damned. Why?

Because a decision maker wants to "know more about me" or because I skipped some social event. Childish reasons like these get me a 50-percent dock on my annual review.

Are you getting the picture now? This goes on every day in corporate America. Many will tell you to adapt to corporate politics and go with the flow. I will entertain their argument. However, if the goal of a company is to maximize profit, how does acting like a child help secure that goal?

I don't say this to whine or cry; it is the way it works at many companies. I checked with other colleagues, and this happens everywhere. I share this with you for two reasons:

1. To give you insight into the motivation behind my loss of interest in corporate politics.
2. To show you what motivated me to build a website of this size. This is the "why."

Getting docked for activities other than performance turned me off from the review process. Once I learned a simple grudge could wipe out my annual review gains, I checked out mentally. I sought advice from a coworker with an incredible judge of character. He told me, "Don, even though you left high school, the games stay the same."

I learned more information from friends in management. Annual review drama is commonplace at corporate workplaces. If one of the "sacred few" has a grudge against you, rest assured your raise and bonus get decimated. Your future advancement is jeopardized until this "problem" gets sorted out. Performance takes a back seat to silly passive-aggression.

I watched other talented coworkers get slaughtered in the review process as well. Enough was enough. I made the decision to never play that game again. I started daylighting immediately. What is daylighting? Allow me to explain.

You may have heard of the term *moonlighting*. Moonlighting is where a person who works a regular job by day takes a second job by night to make extra money. Well, a similar term exists called *daylighting*. CNN coined the term in a 2008 article titled "Sneaky 'Daylighters' Risk Firing by Working Extra Jobs." I define daylighting below:

Daylighting (v)—Working on another project or side business while at your full-time job.

For anybody who asked me how one person builds a site like MathCelebrity with a day job, you have your answer. I present to you .. . daylighting! The next question I get from others in the same situation is, "How do you daylight?" So I share it with you below. If you have any morality hang-ups or ethics snobbery, you may want to skip this part.

I decided to do enough work to make management happy. I served clients, beat deadlines, and kept my head down. Rinse and repeat. I completed the required assignments quickly and quietly. When I finished, I had hours left over in the day to work on MathCelebrity.

So that's what I did for the first five years building the website. Work-from-home days were my favorite because nobody looked over my shoulder. Each day with tactical precision, I completed the work, handed it off for review, and switched over to building my website. I slept like a baby with zero guilt.

Now let me pause here for a moment and tell you: I have zero moral hang-ups or shame about what I did. The company got high productivity out of me and many processes automated. Daylighting produced a hidden benefit for my day job—speed. Because I worked diligently on my website, my programming speed improved. This helped me complete my day-job tasks faster and learn more skills.

For anybody who exclaims, "But you were working on company time," I laugh. Extra productivity gets me what exactly? Not a raise or a higher bonus. It gets me more politics and childish bickering. No, thanks. While the Inner Circle plays its games, I play mine. So long as my productivity is high, nothing else matters.

Some people told me to play the political game better. Friends and coworkers told me I should nuzzle up to management and attend more social events. I argue against this because social events are not listed on my employment contract.

Furthermore, social networking has nothing to do with my performance. I am a firm believer in meritocracies. Call me a dinosaur, it matters not. Why didn't management keep me engaged and discuss social networking? If performance reviews from my peers support paying me more based on merit, why not do it?

Other colleagues were happy with a simple pat on the back or "Good job" from management. They chased the dangling carrot but never quite reached it. Management dangled the promise of potential advancement and the vicious cycle began again. Many talented people got played like fiddles. I learned early in my career never to get caught up in the dangling carrot for a few simple reasons:

• Pats on the back cannot be cashed.
• "Attaboys" do not earn interest.
• "Good jobs" will not buy me shares of stock.

Fluff compliments: That and four dollars gets me a cup of coffee.

The only scoreboard that counts in this game is money. Extra vacation time runs a distant second. Absent management who supports you, the time comes for a person to leave a job. Thirty years ago, people stayed with companies for years. Now, you get out while the getting is good.

When the inevitable day to leave arrived, the same thing always happened. Management asked me to stay. Now the average Joe in corporate America is thrilled with the idea of a counteroffer. The average Joe is wrong, though. When it comes to counteroffers, here is my philosophy as posted on the website Quora:

Rule number one in this game: Never accept counteroffers—ever. Why? Because where was that money before? Now, all of a sudden, management has some magic bag where your extra money just appears? I don't think so. Furthermore, if you do accept their offer, the next year, raise and bonus will be peanuts, and the excuse given will be, "We paid you extra last year." Congratulations, you now have a target on your back.

In addition, never tell your old company where you are leaving to. Let them read it on LinkedIn after you are gone. Do not tell your company details of your reasons for leaving. I'd even give a vague reason for leaving. Let them sweat trying to figure it out. Be polite, but keep your cards close to your chest. As Denzel Washington said in *Training Day*, it's chess, not checkers.

Benefits of this approach include their calling you back in the near future to help them by contracting. If you were a high performer, they will realize the value of their mistake. Because they are now chasing you, you can mark up your rate, you know, on principle. And bonus, now you get two incomes.

This merry-go-round became routine over the years. I started a new job and worked hard for one year. If the first annual review equaled a cost of living increase, I slowed down my pace. If next year's review lacked improvement, I quit my job. The best raises I ever received came from defecting to other companies. If you outperform each year and get tiny increases, you do yourself no favors by staying at a job.

It is the way corporate America works. I used to get mad. When I finally made peace with this in my thirties, things got easier. I focused on building the best and fastest website possible. I treated the day job as a vehicle to pay bills, nothing more. I worked extra hard in the first year to see if the new company rewarded merit. When they avoided performance raises, I reverted to daylighting. So in my second year, I started the "do enough to get by" plan. As long as I satisfied management with progress and no ruffled feathers, nobody got wise to my plan.

No more chasing the carrot for me. I learned to avoid chasing the carrot in my thirties. My dad died of a heart attack when I turned thirteen. He chased the carrot for years as an accountant. Management rewarded my dad's carrot chasing with empty promises and astronomical stress levels. I made a promise to myself to never chase the carrot—ever.

Carrot chasing happened with my colleagues as well. My colleagues told me they went through the same twisted game each year. One carrot-chasing story stood out in my mind over the years. I used to work with a brilliant programmer. Recruiters called him a unicorn. This guy performed at levels most of us dream about. Yet he had his raise withheld one year. Why? For one comment he made in a meeting about another department's productivity. Three hundred and sixty-five days of excellence nullified by one comment! Stories like these confirmed we played a rigged game. Thus, the only extra productivity and extended hours I gave were for my website.

Speaking of productivity, I learned a valuable skill to maximize output. It helped me write thousands of lines of code in shorter time.

CHAPTER 6

MOMENTUM THROUGH CLUSTERING

As I continued building the website, I increased my output speed. I used a concept called clustering to help me skyrocket my output levels. Using linear algebra as an example, I detail my process below:

1. Build a matrix calculation.
2. After completing the calculator, I reused the code to build more lessons.
3. Continue building anything related from the first two points.

One benefit of this process is the ability to reuse code. In programming, one of the important lessons you learn is to avoid "reinventing the wheel." If similar code for your current task exists, you use that code. I used clustering to build linear algebra calculators quickly.

During the two-week period of linear algebra focus, I used one piece of code for all calculators. All linear algebra calculators shared a similar process. This includes the matrix design, the math, and color coding.

I clustered these tasks by expanding on existing processes. Clustering the linear algebra calculators slashed my programming time by 60 percent.

Clustering helped with other math concepts like complex number operations. I note the five complex operation concepts below:

- Add complex numbers
- Subtract complex numbers
- Multiply complex numbers
- Divide complex numbers
- Square root of complex numbers

Each of the complex number operations above has an a + bi term. All five complex operations use similar functions to determine each piece of a complex number. I need to know a, b, and the operation to use. Finally, my code uses the same function for simplifying. Think of clustering as writing a piece of code once to be used many times, with few alterations or updates.

Whenever our fans gave us a new concept for the first time, I built a cluster of tasks for related calculators. This disciplined me to operate in proactive mode instead of reactive mode.

Clustering helped me jump on new problems immediately. As I clustered more problems, the website started to build itself. By reducing tickets, our fans had to wait less. The extra lessons I built through clustering grabbed us more market share on Google.

Clustering gave our website an advantage at the beginning of the school year. If a student ran a certain problem, I searched for all related problems. Why? Because I can predict their next few homework assignments based on what they previously searched for.

Taking this a step further, a few of our loyal fans e-mailed me PDF copies of their books or lesson plans. Some of them sent a syllabus. Now I had a two-week head start on upcoming homework assignments. Math problems share certain properties, so the logic I built can be quickly reused.

Between my old textbooks and the syllabus from students, I had the "what" and "when." I needed the "how," as in, how different students ask for help on our website. At this point, we were a good website getting better by the day. We needed to be great. Learning the "how" took us to a higher level.

ARTIFICIAL
INTELLIGENCE

Spring arrived in March 2010. Angela and I planned a trip to go see my brother-in-law Daniel in Virginia. As much fun as I had building the site, I needed a serious break. Before leaving for vacation, I cleaned up outstanding tasks on the website. A clean slate meant a better vacation.

Shortly before leaving, the game changed for me. On the website, I have a ticket tracker. When a user ran a search on the website with zero results, the website created a ticket. The ticket tracker compiles a list of tickets each day.

The user ran searches with terms like *matrix* or *equation*. The next morning, I reviewed this list and built calculators for the missing terms. Each day, I add more content based on the tickets from the previous day. This process repeats the next day.

Up to this point, the user searched for one word or phrases. Then our entire world changed. Right before my vacation, something interesting happened on the nightly report. It finally appeared, staring me right in the face, row number five on the ticket tracker spreadsheet.

$2x - 9 = 31.$

Somebody entered a math problem! Not a word. Not a phrase. A direct math problem. My jaw dropped. A few rows down on the ticket report, I saw another problem: 987/43. After my initial shock, this made perfect sense to me. Why run a search term or phrase to find a calculator when you can enter your problem directly?

I stared at the ticket tracker report for two minutes straight. Direct math problems symbolized the "how" for MathCelebrity—as in, how do they enter searches? How many ways did people enter these math problems? This presented a big challenge for me. How do I recognize patterns? Building every individual problem cannot be scaled. My excitement turned into disgust. There must be a way to do this.

I woke up the next day and glanced at the ticket report again. I saw more math problems. Just like yesterday, a user entered another long division problem, 425/60. Another user entered a basic math problem, 245 + 38. Another user entered a multiplication problem, 65 × 32. Staring at these problems, a pattern appeared. It took thirty minutes for me to spot it, but it sat on the report in plain sight. Can you spot it below?

- 987/43
- 425 – 60
- 245 + 38
- 65 × 32

Here is the pattern:

1. The term begins with a positive number.
2. This is followed by one of the four basic math operation symbols: plus, minus, times, divide.
3. The term ends with another positive number.

So we have positive number, operation sign, and positive number. With the pattern in hand, I grouped these problems together. I needed one more thing: the sign entered—addition, subtraction, multiplication, division. People asked me how we recognized thousands of problems. I told them the first piece of the puzzle is abstraction. The key is to think in patterns.

I noticed a similar trend reveal itself with fractions.

$$\frac{1}{2} + \frac{3}{4}$$

$$\frac{1}{2} - \frac{3}{4}$$

$$\frac{1}{2} \times \frac{3}{4}$$

$$\frac{1}{2} / \frac{3}{4}$$

Each individual fraction is a number, then a division sign, followed by another number. This is followed by one of the four math operation symbols. The term ends with another fraction. My brain suddenly took interest in patterns. I scanned every ticket report searching for patterns. Over time, I got more adept at spotting them.

As more patterns emerged from the ticket list, I mapped out more shortcuts for users. I also had live examples to build from. I identified two challenges for pattern recognition:

1. Recognizing patterns of math problems
2. Figuring out the math problem pattern a user wanted

I racked my brain trying to quantify pattern recognition. And then it came to me: I needed to solve a problem about solving problems.

If I figured this out, MathCelebrity could become the ultimate math tutor.

CHAPTER 8

REGEX: A MASSIVE UNDERTAKING

I remember losing sleep the morning after the math problem ticket report. The direct math problems consumed my thoughts. I programmed for years, yet I never worked on something like this. The bedside clock showed 2:00 a.m., and my mind kept contemplating. While Angela slept, I sneaked out of bed and ran downstairs to my office. I needed to learn about pattern recognition. I Googled "recognizing string patterns in PHP programming." I read through the first page of articles on Google search results, and one term kept popping up. It remains etched in my brain to this day:

REGEX

Regex stands for "regular expressions." After one hour of cranking through the first page of Google results, I found exactly what I needed. Regex gives you the power to recognize patterns through programming. It also allows you to store these patterns and recall them when needed. The first example I found online had to do with phone numbers.

How do you type a phone number? Without an area code, you type three numbers, followed by a hyphen, followed by four more numbers. But wait, some people may enter seven numbers in a row *without* a hyphen. Or you may have an optional area code added to it. The area code might have a hyphen or parentheses. This thought process turned into a pattern-recognition building map.

- 000-0000
- 1234567
- (708) 1234567
- 708-123-4567

People have a limited number of ways to enter phone numbers. I celebrated this good news. Finite patterns gave me a fighting chance at this project. Yet accounting for more of these patterns created a challenge. The next six weeks began my crash course in everything Regex.

The next day after reading about Regex, I started work on pattern matching. I needed a lesson to practice using pattern matching. In a nod to sentimental history, I chose the first calculator I ever built: two-number basic math operations. Fast forward four hours, and I had a pattern working for addition problems for two numbers.

[0-9]+plus[0-9]+

This initial pattern became the first of hundreds I used on the site. I tested my code update. I ran an addition problem in the search box. The website recognized it and routed it to the addition calculator! I jumped out of my chair and ran around the house celebrating. This little line of code started a new phase in MathCelebrity.

I took pride of the website speed before this update, yet we had potential to be faster. Why search for a math term if a problem exists? It's one less step and gets you right where you need to be. I focused the next phase of the website on pattern recognition.

The week flew by. I built thirty-five patterns for shortcuts during the week on various lessons. It felt like I blinked, and the weekend arrived. That Saturday, my wife and I went out with friends. While we were out, I called my brother-in-law and left him an animated message at 2:15 a.m. The direct math problems piqued my excitement, so I had to share. I had so much fun with the pattern-recognition potential, I had to tell him.

Around this time, a friend asked me about the website. I told him about the latest development: people entering their direct math problems.

He said, "Are you going to hire people to sit and figure out the problems as people enter them?"

I said, "No, I am going to write code to pick up those patterns which then run them through our calculators."

He replied, "So, the computer is going to solve math problems by itself?"

I said, "Yes."

He paused, looked at me, and said, "I say this in a nice way, but that is insane."

I laughed and told him we have surpassed the point of sanity with this project.

I wrote down my notes and set forward with the plan:

• Build as much pattern recognition as possible.

- Review the ticket list each night.
- Build the errors in the ticket report as fast as possible.
- Rinse and repeat.

This worked well for one week. I posted the new updates on social networks. Shortly after, the floodgates opened. Each night for the next two weeks, the ticket report filled up with more than three hundred searches. Many of the errors were direct math problems. My mission to update tickets fell behind. I felt bogged down and I wanted to advance. Just like General George Patton said, "We are never holding our position, we are not holding on to anything. We are constantly advancing."

I reviewed the ticket report one night and noticed something interesting. People asked for the same problem in slightly different ways. Here are example searches with the same meaning:

- 35% of 80
- 35 percent of 80
- 35 pct of 80
- What is 35 percent of 80?

Ah, yes, a definitive pattern started to emerge. First, I needed to update the code to ignore question qualifiers at the beginning of a search. Question qualifiers within searches should be treated as extraneous. Examples are below:

- How much
- What is
- Calculate
- Determine

Besides qualifiers, I also needed to update code to remove common phrase endings. Phrase-ending examples are question marks, periods, equal signs, or any combination of these. The searches below are identical:

- 7 + 8?
- 7 + 8.
- 7 + 8 =
- 7 + 8 = ?

With the qualifiers and fillers out of the way, I progressed with pattern grouping. The percent, decimal, and fraction calculator showed me how people type in their problems. With this strategy in place, the process became clear.

I needed to take the offensive on this pattern-recognition project. To get in front of the ticket barrage, I listed out possible ways people entered a certain type of problem. For example, each search below will have the same answer:

- 3 plus 2
- 3 + 2
- 3 added to 2
- Add 3 and 2

As I studied the ticket report each night, I learned more and more. However, I had to face reality by removing guesswork. Guessing what searches people entered had the same odds as playing the lottery. I needed a better approach. So I grouped patterns to identify user search intent. I let the user guide me from problems they entered. This cleared up confusion and guesswork for me. Now the project captivated me.

I admit when I started the pattern-recognition project, I thought I had bitten off more than I could chew. Identifying direct math problems turned into a big learning curve. Progress started off slow, but I improved over the next few months. I read a linguistics book to improve my ability to recognize language patterns. This painful, yet necessary, process provided the best user experience possible.

In the past, our fans had to enter a word or phrase to find a calculator. As I built this functionality, the website became easier to use. When the word got out about direct math problems, our fans were thrilled. They started entering a truckload of math problems. When our fans saw the improvements, they threw everything they had at the search engine.

I played catch-up for three months. During this time, I worked vigorously on the direct math problem ticket list. I also went back through my math books and listed out the various ways problems were typed out. I set a goal to plow through one chapter a day looking for any possible phrase patterns.

I updated the ticket tracker allowing a person to enter their e-mail for status updates on their ticket. I e-mailed that person when I completed their ticket. Some of our fans used it; others lacked patience. You see, when the word got out about the direct math problem project, some of our fans e-mailed me directly. I list some of the memorable e-mails below:

- Please build this immediately, $2x - 9 = 31$
- Dude, 98 mod 7 is not working.
- I need the sum of the first 50 numbers like yesterday, when will this be ready?
- How do I enter these darn exponents?

Another interesting pattern appeared with various searches. People entered searches in one of three ways:

1. Mathematical notation
2. Typed-out longhand
3. Excel or standard calculator commands

I use a factorial search below as an example. Each of these searches asks for the same thing:

1. 5!
2. 5 factorial
3. FACT(5)

Humans are herd creatures, so this pattern of three made sense. After realizing this, I expanded each pattern-recognition section to account for this. Every time I began work on a new shortcut, I considered the pattern of three if applicable. The ticket tracker gave me valuable insight into human nature. I ran across one constant in human nature: impatience.

Our fans grew impatient with the search engine fixes and e-mailed me their math problems. These e-mails put extra pressure on me to get the search engine working. As the barrage of e-mails and tickets continued, the pressure mounted. As General Patton said, "Pressure makes diamonds." I woke up a few times per week around 3:30 a.m. thinking about these tickets. While finishing the tickets from the day before, I anxiously awaited the next ticket report. Each morning at 5:00 a.m. Central Time, the new pile of tickets arrived.

Building the direct math problem search had a hidden benefit—improving the structure of the calculators. It helped me understand how people entered problems. I structured each calculator to make user entry easier.

Instead of forcing them to enter a problem my way, I let them enter the problem how they wanted. From there, I built code to pick off the necessary pieces of their search to display the math. User experience improved through ease of use.

As I built more pattern recognition, something still gnawed at me. I felt like the website played catch up to user requests. I had a reactive website. I wanted a proactive website. My problem carried over into my sleep. I had a dream one night about running a perpetual race in second place. No matter how fast I ran, I failed to catch the leader.

To solve the problem, I studied more pattern recognition. I read books, listened to podcasts, and watched videos. Unfortunately, continued education provided no solution. So I closed my books, turned off my computer, and stopped thinking about math.

Life provides an interesting lesson in irony. Sometimes, the answer to your question appears in places you never think to look. I finally found my answer in Red Rock Country.

FRACTALS, FORUMS, FRESH AIR

Mother Nature never ceases to amaze me. If you appreciate nature's glory and feel like taking a trip, block off eight hours of your day. Start in Phoenix, Arizona, get in your car, jump on I-17, and drive straight north for 115 miles. You'll run into a small town called Sedona. Red rocks, clean air, and breathtaking views dominate the landscape. The hiking trails clear your head and cleanse your soul. No matter what problems life brings you, a day in Sedona will change your outlook.

Four years after starting MathCelebrity, my wife and I took a vacation to Arizona. We spent a full day in Sedona. What I saw in the little Arizona town changed my outlook on math. Deep in the hiking trails, facing red rock views, I found a new way to solve math problems. Mother Nature educated me on Human Nature.

The solution to my problem is something we see everyday. Seashells, snowflakes, shorelines. All three represent a phenomenon called fractals. Fractals are geometric figures found in nature. Each smaller part of a fractal mimics the whole. In the Sedona mountains, you'll find fractals staring back at you. Rock formations and tree branches paint a fractal picture. As I stared at the rocks and trees, it came to me. I struggled to quantify it at first. As I spent more time on the trail, I couldn't ignore it. As the sun set on the gorgeous sandstone formation, I found my solution.

Math problem variations are fractals! You have required pieces, and you have optional pieces. For optional pieces, they must contain the original required pattern. Whether it's unknown equations or long division problems, their identities represent patterns.

First, let's look at some examples:

$x = 9 \rightarrow$ Required variable, required equal sign, required number after the equal sign
$x + 3 = 9 \rightarrow$ Add an optional number
$2x + 3 = 9 \rightarrow$ Optional Coefficient
$2x - 3 = 9 \rightarrow$ Subtract an optional number
$2x + 3 = -9 \rightarrow$ Optional negative sign on the right side of the equation
$2x - 3 = -9 \rightarrow$ Subtract an optional number

Go through each equation or inequality and find what they have in common. I start with one variable equal to a constant. From there, I create similar equations by adding coefficients and extra operators. Finally, the equal sign operator may be an inequality.

Determining required pieces for an equation or inequality is key. We have a variable. We have an operator. We have a positive or negative number on the right side of the operator. From there, we expand to other versions of problems. All versions contain the three required pieces.

Now let's go deeper down the rabbit hole. I listed positive or negative integers in the example above. Let's expand our universe and get inside the user's head. What else might they enter?

What about decimals?

- $x = 9.25$
- $2.5x = 10$
- $3.5x = 10.5$

Don't forget about fractions!

- $x = \dfrac{1}{2}$

- $\dfrac{2}{3}x = 10$

- $\dfrac{3}{5}x = \dfrac{10}{8}$

Let's think about fractals again using trees as an example. The tree trunk represents the required pieces in each problem iteration. The branches and leaves connect to the tree trunk. Each branch represents different combinations of the optional iterations. Remember, branches are still related to the tree. This relational pattern is what I saw in Sedona. The MathCelebrity search engine is a fractal. Visualize a tree, where math problems are the trunk, and each branch is a calculator. My goal is determining how to construct each branch.

Thinking in fractal patterns, it helps to cut out what you don't need. Let's use my tally mark calculator as an example. Tallies are lines or marks representing a number. ||| is 3.

1. We want a positive number. It's never negative. It's never a decimal. It's never a fraction. So $\{1, 2, 3, \dots\}$
2. We want the word tally, tallies, or tally marks immediately after this number.

If the search engine detects this pattern, it knows this is a tally calculation. The search engine requires the first and second piece above. If it lacks the number and tally phrase, or it has extra information, it's not a tally calculation.

Think of a calculator as a tree. Now, each branch is a shortcut attached to the tree. Let's use tally marks as an example. The tally mark calculator is a tree. The shortcuts below represent branches on the tree.

1. 52 tally
2. 52 tallies
3. 52 tally marks

Each calculator contains unique patterns. No two trees are alike, and no two MathCelebrity calculators are alike.

I came across one more challenge using fractal thinking: similar patterns between problems. Consider the searches below:

- (1,2)
- 1,2
- {1,2}
- (1,2,3)

The first search could be a two-dimensional point on a graph, or it may be a two element number set. The second search is a point or a number set as well. The third search contains braces denoting a number set instead of a point. The fourth search may be a three-dimensional point or a number set.

Fractal thinking created an advantage for me. You see, once a unique problem or concept appeared on the ticket report, I had a blueprint. This blueprint gave me other problems to start building. Remember in the clustering chapter when I built calculators in advance? The fractal method built my patterns in advance. If somebody wants to run a math problem, my website needs two things. The first is a calculator to display the math. The second is a pattern to identify what type of problem they want to run. To describe the relation: clustering is to calculators as fractals are to patterns.

With a fractal frame of mind, I turned the search engine into a proactive machine. All I needed was a problem. After defining the problem, I determined the possible patterns. Each pattern has required and optional inputs. I determined the required pieces first. Next, I determined what was optional and also replicated the pattern. To recognize a pattern, I needed to know what to eliminate.

I treated the fractal approach as a puzzle piece. Once I had the first piece, I imagined different ways a user could enter similar problems. Each additional pattern represented another puzzle piece. When you build a puzzle, you look for the correct pattern of the next piece. Puzzle solving also requires you to identify mismatches. This led me to step two of my revelation in the mountains: mismatches.

Using a fractal approach, I wanted to eliminate extraneous information. I always remembered the old quote from the Renaissance artist Michelangelo. When asked how he sculpted the statue of David, Michelangelo replied, "It is easy. You just chip away the stone that doesn't look like David."

Remember the tally calculator example above? I looked for a number and one of the phrases with the word tally in it. If I saw anything else, I discarded it and moved on.

I almost forgot to mention the last piece of the puzzle: people. Since people think in fractal terms, I wanted to understand their thought process. To learn more about people, I created a forum. I built the forum for three reasons:

1. Bring people together
2. Open up the discussion
3. Learn how people ask questions

I wanted to bring people together by giving them a sense of teamwork. The forum had one distinct advantage over the calculators: transparency. When somebody ran a problem through the search engine, nobody else knew about it; searches remained private. While searches were private, forum questions were public.

Opening up the discussion stimulated peer validation. Since people look for peer validation, more posts created more discussion. When I first built the forum, I pushed the discussion forward. I posted fifty word problems over the course of five days with full solutions. I categorized each post similar to search engine calculators. Nobody signed up at first. As I kept posting, members started trickling in. I posted word problems from our support tickets. After I created 100 posts, members began asking questions on the forum.

The forum had trending topics, recent questions, and active posts. Each post had a time and date tag to display recency. Popular threads moved the discussion forward. We tracked post views; more views meant higher popularity. Clear categories provided easy organization.

The forum questions improved calculators. The calculator searches created more forum posts. Many times in the forum posts, part of the solution linked back to the calculators. The website grew more branches. Since forums are self-organizing, more patterns emerged.

One such pattern appeared each year: seasonality. Certain types of questions appeared at certain times of the year. During finals, confidence interval and hypothesis testing questions flooded the website. Near the beginning of the year, word problems such as coin combinations popped up. Distance problems appeared in groups as well. Trending forum topics showed current homework assignments across the country. Forum insights told me a story about our users.

Posting time and view counts provided extra insight. If students procrastinated on homework, questions popped up in the morning before class started. Morning questions contained more urgency in the comments. When students did homework at night, the posting time painted the picture.

Armed with this new knowledge, I expanded the search engine. People created problems, problems turned into patterns, and patterns drove progress. I found my breakthrough using fractals. The entire MathCelebrity search engine is a fractal. This fractal consists of smaller fractals. This fractal approach helped build the search engine into a powerful pattern recognition tool.

As the search engine grew, the code turned into a monster. We suffered from code bloat. The amount of code I wrote for shortcuts slowed down the search engine. Every second lost on our search time damaged our reputation. I needed to improve this immediately. MathCelebrity needed a tune up.

THE QUICK
AND THE DEAD

I read an interesting statistic about website speed. It said if a website takes more than two seconds to load, users will get frustrated and leave. Around this time, I read more articles discussing faster websites getting search-rank boosts. Google led this charge with its search algorithm updates. The timing coincided perfectly with a site speed audit I worked on.

With the large amount of direct math updates I added in the code, our speed dropped. I tested the site to measure page load times. I gave our speed a B+, yet I wanted an A. To meet this goal, I needed to eliminate the bottlenecks in the code. After four hours of investigation, I found the low-hanging fruit. This file controls our search logic, which I refer to as our "hostess." Who is the hostess, you ask? A better question is, what is the hostess? The hostess is one of the secrets of MathCelebrity. I've kept it secret for eight years, and now it's time to share.

I often get asked, "How in the world did you build this website?" I structured MathCelebrity as a restaurant waitstaff. Stay with me, I'll explain and introduce you to the hostess and the rest of the staff. First things first, let me welcome you to the MathCelebrity restaurant:

1. Imagine you walk into a restaurant.
2. A hostess greets you and asks you if you have a seating preference (window, near the bar, etc.).
3. Next, you open the menu and order (appetizer, main course, dessert).
4. Your menu choice determines what table you are seated at.
5. Each table serves only one course.

Now, I want you to meet the MathCelebrity staff and learn each of their roles:

Restaurant Employee	MathCelebrity Job
Hostess	Search engine determining what you want
Table	Each individual calculator with one job only
General Manager	Backup for the hostess if multiple answers to a question exist

I have one piece of code with three thousand lines called the hostess. It determines what you are asking for. The hostess avoids math tutoring, solving, or detail. Her job is simple. She takes your request and routes you to the appropriate table in the restaurant. Have you ever walked into a restaurant and the hostess looks into the computer to check the seating chart? This is how the MathCelebrity hostess operates. Let's compare a restaurant with MathCelebrity.

- Are you searching for a word like *equation*? If so, go to the main seating area and I will show you everything we have related to your "meal request."
- Did you enter an exact problem, such as $2x - 9 = 31$? If so, the hostess recognizes you entered an equation. She sends you to a specific table

dedicated to solving equations. Call it the equation table. All you get is equations all day every day. It is a separate piece of code focused on one task with ruthless, unapologetic focus.

- Did you enter a problem or term with multiple interpretations? It's time to meet our general manager.

Each table in the restaurant is a calculator. At the time of this book, we have 450 calculators. The hostess interprets your search request and directs you to the appropriate table. The general manager is called if multiple choices or interpretations exist.

This way, each table is called on to solve a particular type of problem. It has one job and one job only. It never interferes with other tables. For instance, the matrix table never interferes with the equation table. The matrix table solves matrix problems only. Each calculator has one job. One job means simplicity. Simplicity leads to a lightning fast website.

Each table is an algorithm. For those who are wondering, an algorithm is a fancy word for a set of steps. These steps solve a problem and produce a final product. A simple example of an algorithm is a recipe to bake a cake. You follow a set of steps to get an end result.

1. Beat two eggs in a large bowl.
2. Add one stick of butter.
3. Add two cups of sugar.
4. Add one cup of flour.
5. Mix all of the ingredients above in a large mixing bowl.
6. Pour the batter into the prepared pan.
7. Bake in oven on 350° for thirty minutes.
8. After thirty minutes, take the pan out of the oven and enjoy your cake.

Let's return to our restaurant example for MathCelebrity. I will use our even-odd calculator as an example. The even-odd calculator is a simple algorithm which I detail below:

1. Enter an integer we will call (n).
2. If n is not an integer, stop and send a message to the user
3. If n is an integer, continue below
4. Divide n by 2 $\Rightarrow \frac{n}{2}$.
5. If no remainder exists, the number is even.
6. If a remainder exists, the number is odd.

The even-odd (table) has one job: to determine if a number is even or odd. The table produces the step-by-step math work behind it. Let's run through a typical request.

A customer comes into the MathCelebrity restaurant and asks, "Is 5 an even number?" The hostess recognizes this direct request is an even-odd question. The hostess routes this request to the even-odd table. As she walks the customer over and sits him down at the even-odd table, she tells the table employee "5." The table takes the number 5, runs the steps above, and determines the number is odd. The table shows why, since 5 divided by 2 has a remainder of 1.

Let's pause for a moment. Notice how the hostess gave the table employee only one number? Why? Because the table runs only even-odd calculations. The even-odd table employee only cares about one number.

Now back to the speed improvement I referenced earlier. The hostess file had additional potential for speed. I found other decisions in the program to combine or eliminate. I sliced and diced the code to improve the search speed.

I finished cleaning up the code over a three-day period. Two of the fixes resulted in a 15-percent faster website. Unfortunately, the next fix I made broke the site due to poor testing. Luckily, a fan alerted me to the mistake shortly after I broke the website. I corrected my mistake and improved the hostess file even more. By the end of the next day, the website gained more speed. That concluded the changes with the hostess file.

I forgot to introduce one more key player in this scenario—the general manager (GM). Have you ever been to a restaurant and had a question or problem you needed to discuss? At a restaurant, the GM surveys the floor answering questions and greeting guests. Food problems? The GM will straighten it out. Employee issues? The GM irons out the details.

The GM for MathCelebrity provided oversight for the hostess. When you enter a term in the search box without a definite answer, the GM steps in. For example, say you enter the term *matrix*. MathCelebrity has a few calculators related to the matrix. Because the hostess's job is to find definitive answers, she passes this request to the GM. The GM determines this is a word-based search and presents all calculators related to a matrix. The GM presents this list on a screen in the search results. When the customer chooses an option, the GM tells the hostess what table to go to. A direct request is determined, and the appropriate table takes it from here.

What about math problems with multiple interpretations? A fraction, for instance, such as $\frac{3}{4}$? The GM comes to the rescue again with interpretation skills.

The GM determines all available options for this problem. Fractions can be simplified, converted to a decimal, percentage, or a unit fraction. The GM steps in to present ALL these options to the user.

The user selects their option, and they are routed to the calculator of their choice.

Using this approach, the restaurant functions efficiently, and speed is increased. Now you know why our searches run in one second or less—isolation. Everything is put in separate compartments which isolates each calculator. Just like you see on submarines, compartments provide a safe, efficient structure. There you have it—the big secret behind MathCelebrity.

To fix the hostess file and make it faster, I created groupings. Groupings are shortcuts for similar math problems. The problem with the hostess file related to rewritten code and reinventing the wheel. To correct this, I created a grid of calculators with similar shortcuts. I grouped the similar shortcuts into a reusable section of code.

Let's say ten of our calculators had similar shortcuts. Instead of writing shortcuts for ten different files, I used one block of code. This block of code handled any shortcuts for the ten calculators. After five days of rewriting the hostess file, I updated the website. It shaved the amount of code in the hostess file by 25 percent. It also cranked up our speed by 35 percent. Not only did this make the website faster, but groupings also simplified new shortcuts.

An example of grouping is measurements such as ounces, inches, and seconds. Our users asked for a number and a measurement. They typed a positive number followed by the measurement they wanted. The measurement contained a singular or plural format. Examples are below:

- 8 ounces
- Convert 4 inches
- 20 seconds =

This is one of more than fifty groupings on our hostess file. One small code block handles all types of measurement problems. I am proud of the speed improvement gained by groupings. Our fans showed immediate approval. After concluding the speed updates, I moved on to other features.

CHAPTER 11

QUIZZES AND OTHER FEATURES

A few parents wrote in one day concerning quizzes and tests. Their kids struggled on live tests but found benefits from extra practice. The two forms of practice they craved were practice problems and practice tests. The parents said timed tests helped the kids track progress and get faster. Another lightbulb went off—I should build quizzes and practice problems on the website.

I began with the easier of the two tasks, practice problems. I jotted down ideas and decided to add an extra button on each calculator. The button read "Generate Practice Problem." When a user pressed it, the calculator generated a random problem. At the time, the website contained more than four hundred calculators. I used my tried-and-true method expanding the website: build one calculator update and use it for all other calculators.

For all applicable calculators, I took each piece of a particular problem and randomized it. Take something simple such as the product of binomials. We always have $(ax + b)(cx + d)$. In this particular sample problem, we have the following pieces:

- One variable (x): We randomize this with any letter a–z except for e and i which are special letters in math.
- Two coefficients for variables (a and c): We can use any positive or negative integer except for zero.
- Two constants (b and d): We can use any positive or negative integer except for zero.

The practice-problem generator randomizes the variables and constants, pieces the problem back together, and presents the practice problem to the user. This method generates a different problem each time. You get unlimited practice not found in a textbook. Parents loved this feature because it spared them hours from dreaming up random problems. If you want to win the hearts of moms with kids in school, find a way to give them more time back.

Besides time savings, infinite practice problems won us praise. Other practice problem generators limit your practice. They have a finite number of practice problems. Once you run through their problem set, everything repeats. Over time, it gets stale. On the contrary, MathCelebrity gives you unlimited practice. One parent at an education conference called this feature "drill and kill." Think of it as practice until you can do it with your eyes closed.

Here is the best part: after generating your practice problem, all you need to do is press the Calculate button to see the step-by-step solution to your practice problem. Press one button for practice; press another button to calculate. Both buttons are located next to each other. Fast and easy is our style.

Now, on to the quiz generators. I built our quiz generators using another algorithm:

- Determine all randomized pieces like we did with practice-problem generators.
- Determine any special parameters for the quiz, such as having only whole number answers.
- Gather up the final questions, answers, correct answers, and time taken.
- Grade the quiz.
- Time the quiz.
- Show a green check mark if the user answered correctly.
- For each wrong answer, show a red "X" with a link. This link displayed the correct answer with the step-by-step solution.

The quiz generator paid dividends in the form of scalability. A teacher gave the same quiz to thirty different students and none of them had the same questions and answers. Student exams had the same concept and problem type, yet randomization made cheating impossible. The parents and charter school teachers we spoke with loved this feature. Who has time to make thirty different random quizzes and answer keys? That's what MathCelebrity is here for.

As the old saying goes, no good deed goes unpunished. We received more requests on the quiz generators. Parents wanted to save quiz scores to their personal account. This feature notified parents of the quiz score, date, and time. Parents shared log-in accounts with their kids. These e-mail accounts controlled log-ins and notifications, so I set this up next. Personalized accounts granted parents power to control problem randomization.

For example, a student starting out in basic math adds small numbers. The parent of this child sets a maximum and minimum number for addition. Let's use a range of 1 through 100. This student only sees small number addition up through 100. Custom randomization allows parents to adjust problem level difficulty. More advanced students may set top-end addition numbers to 10,000.

MathCelebrity stood out from the competition using randomization. Other education websites have a limited number of practice and quiz problems. After you take the quiz a few times, you start to see the same problems over and over again. In contrast, the random-number generators used on MathCelebrity always shuffled the math problems. Custom randomization sweetened the deal

With practice problems and quizzes completed, I shifted focus to memory improvement. I added flashcards to our premium plan next. Math and memory go hand in hand, so this feature improved the website experience. Formulas and shortcuts are important to your math memory. I structured this feature like a practice problem with an animated flashcard effect. When you clicked the flashcard, it flipped over and displayed the answer. When you clicked it again, it displayed the original question. The animation provided a nice interactive feature.

Next, I improved calculator categorization, organization, and relation. Some of our users like to start by browsing concepts. Alphabetical search is a common browsing technique. I built a two-step alphabetical search.

At the top of the page, we displayed the letters *A* through *Z* as links. When you clicked a letter, it showed every concept we had beginning with that letter. Each concept contained a clickable link. This link displayed all calculators related to that concept. We now had a math dictionary. The dictionary grew larger with every concept we added.

I built a subject search for our next organizational improvement. However, differentiating subjects by state created a challenge. Certain states treat a concept as pre-algebra while other states classify it as algebra. To set a standard, I used Illinois for our state subject hierarchy to display my lessons.

Those changes gave us categorization and organization. Now we needed relation. To accomplish this, I inserted related key word links at the bottom of each calculator. Let's say you were on the circle calculator. One of the key words on the circle calculator is *radius*. When you clicked this link, the search engine returned all lessons related to the word *radius*. This structure helped people find information fast by linking related concepts to one another. Visualize for a moment, a chain link. Each link ties to the next. Categorization, organization, and relation built a chain link made of math.

After I built these features, I added the ability to save specific calculations to an account. A student or parent bookmarked their calculation during a homework session. They returned to the saved calculation on their account by clicking the link. The link reran their calculation without them having to retype it on the search engine.

With these extra features, MathCelebrity became a powerful resource. Yet something still nagged at me. I never gave an in-person demonstration of these features. While e-mail feedback is great, nothing replaces face-to-face interaction. Thinking about this begged the question, whom should we give a demo to?

CHAPTER 12

MEET THE PUBLIC

After five years of building this mathematical mammoth, my wife and I decided to promote the website. I had fun for five years building this beast in my basement. Yet I classified this website as a hobby, not a business. A business earns money and a business grows. While I worked hard on this project, I only had thousands of visitors to show for it. The time arrived to monetize the website.

After researching forums online for high-traffic websites with no premium products, three things came up:

1. Advertising through Google AdSense
2. Direct advertising for businesses
3. Affiliate marketing

I started with Google AdSense due to ease of use. You create ads by installing the code on your website in less than two minutes. Google takes your content, figures out the preferences of your user, and serves up an ad. I placed three ads, the maximum allowed by Google, on each page of the website. In less than five minutes, you start earning money on your website.

Google AdSense pays you when a person clicks your ad or when an ad shows one thousand times. Google made this process idiot-proof. We made money the first few months. I think our first month topped $100. Nothing to write home about, but it's automated and easy money. When businesses advertised on our website, they received a large amount of exposure. Our advertisers benefited from the amount of page views and time spent on our website by our users. This gave them more brand exposure.

Direct advertising for businesses became our second monetization option. A few people got in touch with me and asked for a quote to place banner ads on our website. Because we owned valuable website space, I declined cheap offers. We needed more traffic to attract high-end media buyers. My friend David Kimrey told me one million unique monthly visitors will attract these advertisers. I wrote this down as a future goal to strive for.

Affiliate marketing provided another monetization strategy. Here is how it works: You promote goods and services on your website. If a buyer purchases from your affiliate link, you get paid a commission. Amazon is one way to earn affiliate commissions. You promote products on Amazon. If website visitors click your Amazon ad and buy within twenty-four hours, you get affiliate commission on ALL goods they buy. I explored a few other options and decided to stick with AdSense first.

I used AdSense for a few months and it made a little bit of money. During those few months, something nagged me. I felt the website delivered serious value, but we were limiting our revenue growth. I thought of other ways to monetize the website.

We needed to get serious with our website and treat it like a business. I secured a trademark for the name MathCelebrity in the education domain. My wife and her friend who is strong in graphic design started working on the logo. The logo came out perfect and represented our brand well.

My wife and I discussed the future of the site. We needed to take MathCelebrity to the next level. Sitting in the basement writing thousands of lines of code will not move this business forward. Advertisements alone will not propel us to the next level. Angela and I decided to sign up as an exhibitor at an education conference.

Two educational shifts supported our decision. Homeschooling numbers in America took a vertical climb and education shifted online. As I wrote this, America had 1.5 million homeschooled children. Digital schools and online learning programs popped up like weeds. The digital education market had white-hot potential. We needed to capitalize by going to this conference.

The conference gave us a chance to interact with people who needed math help. I tailored my plan based on the crowd response. If people gave us a strong response, we had justification to charge people. If they were indifferent about our website, we continued with Google AdSense. With strategy set, we shifted focus to our venue. After researching educational organizations, we found an ideal match. We booked the National Council of Teachers of Mathematics Conference (NCTM) in October 2010.

After booking the conference, I had insomnia. I laid awake thinking about the outcome. Sure, we had some decent success with online traffic up to this point. Yet we had not demonstrated this product in person. I put in thousands of hours on this website so far. Preparation is great, yet three questions kept me awake at night:

1. Will people care?
2. Would they understand the website?

3. Are we going to flop in person?

I decided to bring my wife and my brother-in-law Daniel along to help talk to people. My wife and brother-in-law have a warm, approachable demeanor. They were the perfect choice to attract guests and give demonstrations. My role consisted of answering questions and telling our origin story. If you watch superhero movies, you know how important the origin story is. People want to know about more than your product before they hand over their money. People buy people more than people buy products.

With our action plan set, my team and I flew out to Baltimore. We had three people, two days, and one mission.

The conference started at 8:30 a.m. Things moved slowly for the first thirty minutes. The first few people who passed our booth were in browse-only mode. At 9:15 a.m., we spoke with our first conference attendee. My wife and brother-in-law worked on their script and approach. After speaking with a few more people, they hit their stride and their presentation flowed.

As more guests stopped by our booth, the feedback we received floored us. We heard the same five phrases throughout the conference:

- I cannot believe something like this exists!
- I have never heard of you guys before; this is incredible.
- Where was this when I was in high school?
- WHY in the world is this free?
- This is a lifesaver!

Teachers and parents noticed the detail level of the math work. A few commented on how methodical the step-by-step work looked. This is how I differentiated our website from other websites. Many websites may show an answer or only part of the math. We show the interim math. Here's an example:

- $2x - 9 = 31$
- Add 9 to both sides.
- $2x + 9 - 9 = 31 + 9$
- Cancel the 9 on the left side, we get:
- $2x = 40$
- Divide each side of the equation by 2.

- $$\frac{2x}{2} = \frac{40}{2}$$

- Cancel the 2 on the left side to isolate x.
- $x = 20$

The math work detail shown above helped us stand out at the conference. I avoid fill in the blanks with the supporting work. Everything is laid out with lucid detail. You only see the answer after scrolling through the step-by-step math work. This core benefit helped us stand out against our competition. Parents liked seeing the work first leading up to the answer. This early feedback gave us momentum leading up to lunch.

All of us huddled during our lunch break to discuss the conference. Our jaws hit the floor; people loved the website. What an incredible feeling! We even had other vendors come over to our booth to ask about our website. The vendors told us customers came to their booth raving about our website. The vendors came to see what got people so excited.

My wife and I came to the conference with two goals—introducing the website and getting feedback. We had no other expectations from guests. We hoped to make a good impression and come away with some valuable insights. By lunchtime, the positive momentum engulfed us.

I remember taking a quick bathroom break and forcing myself to wipe the stupid grin off my face. Moments like this make you feel like you can walk on water. The three of us awaited the opportunity to meet more people.

There were things to improve. People asked for certain problems the website did not handle at the time. The pattern recognition logic needed a tune up. Conference attendees judged us on our problem-handling ability. I had a choice: Wait until after the conference, or make the fixes right in the booth.

I had so much energy from the conference so I coded the fixes right in the booth. I wandered through the aisles to find the people who requested the fix. When I tracked them down, I fixed their problem. I asked them to run it again to confirm. For the people I missed following up with, I sent them an e-mail detailing the fixes I made.

Teachers had a few questions for us:

- Nobody had ever heard of us before; who were we?
- Other established programs were out there; how did we get started?
- Teachers used Khan Academy which is very popular; how were we different?

Unfortunately, I had weak sales skills, so I lacked a crafted message for this. I needed to fix this sooner rather than later. I created a list of improvements to make from the feedback we received. We ended the first day of the conference with a clear mission. I closed out day one with two objectives: improve the calculators and clarify the website benefits.

The conference ended at 5:00 p.m., and Angela, Daniel, and I went out to dinner. We discussed four things:

1. We have a product people want.
2. We should start charging—immediately.
3. We should keep making continuous improvements.
4. We should go to more conferences.

Day two escalated into another emotional high. We met new teachers and reconnected with contacts from day one. The people from day one returned to our booth and asked us more questions. From these discussions, we learned a ton of helpful information. For example, we received valuable insights about our competition. We learned about their products, services, and popularity. A few of our competitors were established companies. I welcomed competition: it makes you turn up your game. This is what I like to call good pressure.

We received a valuable piece of intelligence on our competition. Teachers had certain perceptions of our competitors. We found out which ones they liked. We also found out what they wished our competitors had. Conference attendees told us our speed and detailed math work gave us an advantage. Our sheer volume of calculators and concepts gave us another advantage. Besides competition, your colleagues provide valuable intelligence.

Later that night, we met with some of the vendors at the conference. I learned a valuable lesson: The reaction of your competition and colleagues tells you how well you are doing. The conversations we had with other vendors surprised us. We were humble, and we knew we had improvements to make. How the vendors perceived us remained a mystery.

When we met the vendors, we heard great news. The vendors spoke to us with admiration and praise. It felt great. They kept talking about the teachers who swarmed around our booth. They told us how people stopped by their booth and raved about our website. These conversations boosted our morale.

I remember several vendors dropping by our booth during the conference. I thought they wanted to be friendly, but I realized how naïve I acted back then. The vendors stopped by to learn more about us. From the conversations with the vendors after the conference, I knew we were on to something big.

We closed out the conference with a long list of e-mail contacts to follow up with. My wife compiled a list of feedback from teachers. Educators wanted to stay in touch with us and receive updates. The conference exceeded every expectation I had. We left the conference with a clear mission and improvements to make. I hoped this positive outlook continued for the weeks ahead.

CHAPTER 13

POSTCONFERENCE HIGHS AND LOWS

We flew back from Baltimore, and I wasted no time improving the website. I followed up immediately with people we met at the conference. I updated the website based on the feedback given. I built more calculators, shortcuts, and detailed instructions for the math work. I wanted to cover as many searches as possible, leaving nothing to chance.

Next, I got to work on a pricing structure. I updated the pricing three times over the next year. I decided to charge for a license without any free trials. Let me pause here for a moment in case you have not figured it out: my marketing skills were poor. I put a new process in place for a premium plan. A person runs a calculation, and the website redirects them to a bland sales page. This page had a quick video summary explaining why the service requires payment.

I also crafted a solid answer for the Khan Academy comparison raised in Baltimore. Let's use a long division problem as an example. With Khan Academy, you watch one video example of a long division problem. What if you have a new long division problem with different numbers?

Khan Academy does not have another video for your new problem. You have to replace the numbers from the original video with your new numbers. In contrast, MathCelebrity solved ANY long division problem.

Our speed gave us another distinct advantage. Who can beat one second or less? Our strengths did not stop there. We had other serious advantages over other educational websites and tutors:

- There's no need to watch an instructional video.
- There's no need to consult a boring textbook.
- There's no need to leave the house to be tutored.
- There's no need to spend a pile of money.
- The student controlled the pace of the learning.
- You get what you need instantly.

Students e-mailed us about how much time they saved on homework using MathCelebrity. When a student saved thirty minutes to one hour a few times per week, it adds up fast. We used these time-saving numbers as a bold testimonial. With the positive feedback we received, all roads led back to the search box.

The search box simplified the entire user experience for our fans. The search box held their hand and gave them anything they wanted. While crafting the sales message, I came up with the nickname "Google for Math." When people think of Google's rectangular search box, they think of simplicity. It gets them what they want immediately. This phrase stuck, so I ran with it. From there, I added a few revisions to the initial sales message with the testimonials. With our testimonials and sales pitch in place, the time came for a premium service. I flipped the switch turning MathCelebrity into a paid website.

During the first six months, the membership plan flopped. My messy sales pitch and the lack of free trials hurt our sales. Some people bought the membership, but the conversion rate dipped below 1 percent. What a sad number. A few people wrote in to tell us they don't pay for anything on the Internet. Others cursed us out with disgust about us charging. I paid them no mind. I pressed on for nine months like a stubborn bull. Nine months gave us time for the freebie seekers to drop off. I expected traffic to decrease, which it did.

Shortly after updating the pricing structure, we went to a homeschool conference in California. Armed with a better sales pitch, my team captured more leads and handled more objections. Our sales pitch generated more memorable interactions with people at our booth. One of these interactions turned into a testimonial we use to this day. I share this story any chance I get.

A family came up to our booth to ask about our service. My brother-in-law began his presentation. While he spoke to the mom and dad, their fourteen-year-old son walked over to one of our laptops. He started playing around on the website. He asked my wife a few questions about how the site worked. After giving him a brief introduction, my wife showed him the matrix operations calculator. This calculator is her favorite because she loves the color coding. The color coding impresses users, so we make sure to feature it in our demos. After she gave her demo, she let the son try out the calculator. After ten minutes using the website, the son said, "Mom and Dad, you have to see this. I know how to multiply matrices."

Mom and Dad walk over to the laptop. My team followed the parents with breathless anticipation. I noticed the son using the Matrix Operations Calculator. He explained to his parents how he learned matrix multiplication using our calculator. He liked the color coding and the step-by-step math work. Impressed, his mom said, "This is material he will need to know in college."

I kept smiling after the family left our booth. What an incredible feeling to know our calculator taught a tough lesson in a short time. Later on, this family brought over other families to check out our website. The other families told us they had to check out the website after hearing from the parents. Even though this happened a few years ago, I remember it like yesterday.

The matrix calculator story stuck in my head during the conference. A high-school kid learned college concepts in ten minutes. He didn't even step foot on a college campus, yet he started learning college math. Then it hit me: we had another major benefit to present. Students could jump forward or backward between lessons. They were not confined to a grade level. This proved to be a unique benefit of our website. Many educational websites force you to choose a grade level to complete. They block you from skipping ahead to other grade levels. With MathCelebrity, you navigated where you wanted, when you wanted, hassle free. Our sales message continued to craft itself.

As the conference went on, we met more people, some of whom discussed partnering with us. Another good sign for our future. Charter schools, educational groups, and even potential affiliates talked about working with us. Our list of contacts grew with each conference.

As the conference ended, I looked back on what we learned. We had a successful conference. We sold licenses, we made contacts, and people gave us great feedback. Like the last conference, we had other vendors coming over to see our website. One vendor told us she wished the conference put her booth near us since people flocked to see our website. We heard one of our favorite pieces of feedback many times during the conference: "Where was this when I was in school?"

Yet the situation had a darker side. After the conference, our sales online were grim. While we sold licenses at the conference, our sales pitch needed improvement. Our follow-up marketing needed work. The years I spent developing the website came with a cost: I neglected sales and marketing. As the old saying goes, "Nothing happens until somebody sells something." The time came to make a decision on the future of the website.

After three more months of this madness, I pulled the plug. I needed to regroup and learn what went wrong. I took notes on purchases and feedback during the first year we charged. During this year, we made a small amount of revenue.

After this experiment, I decided to make the website free again. It took two weeks before traffic started roaring back. Fans expressed excitement when we switched back to a free website. Looking back on this period, I think we made a mistake not giving free trials. Our poor sales message needed improvement as well.

I charged a low price for a service people loved. Why weren't they buying? Part of me thought we caught the "freemium virus," where people expect everything for free online. I wrote it off as people, especially younger people, wanted everything for free. The trend online seemed to support it. Some examples of people not paying for products and services online include the following:

- People pirating music and other digital content
- People bootlegging e-books instead of paying for a copy
- People abandoning websites once they figure out a product costs money

I read forum posts for various products where the first thing people ask is, "Where can I get this for free?" Did the younger generation expect constant free stuff? Part of me still thought so, but I lacked evidence to support my theory. One theory crossed my mind: we may have been free for too long. Because of this, people got used to free. I grew tired of speculating. I flipped the switch to make MathCelebrity free once again.

Two weeks after returning to a free website, I came across a few marketing blogs. I spoke with a few people involved with software as a service (SaaS) products as well. I learned some valuable tidbits in that time. My entire mode of thinking began to change. I needed to improve my follow-up with potential customers.

People loved our service. Their feedback on our track record confirmed it. I needed to educate our fans by showing them the value of our service. I did this by following up on e-mail with our fans. To grow my skill set, I began my education with marketing courses.

CHAPTER 14

MARKETING CRASH COURSE

I blocked off time to learn more about marketing and sales. I found Dan Kennedy online and began reading some of his material. Dan is the teacher I wish I found five years ago. For those of you who don't know Dan Kennedy, he is one of the premier information marketers out there. He makes a pile of money giving you necessary advice he learned in the marketing trenches. His nickname is the Professor of Harsh Reality. After listening to some free audio conferences and reading his newsletter, I got hooked. I bought all his books and began my marketing journey.

I learned a few core marketing lessons during this time. Sadly, I violated all of them. Lesson one: capture leads to e-mail and call them. But I needed something to offer them to demonstrate value. I created an eighty-page study guide of math shortcuts and formulas as a free gift. Next, I set up an opt-in form to collect contact information. When a person filled out their first name and e-mail, I sent them a PDF study guide.

Lesson two: give before you receive. I followed up with fans who signed up giving tips, tricks, stories, and other useful math knowledge. I love to write, so this part came easily to me. I tweaked the approach a few times, but click-through rates started going up. We had a few people each day signing up for our list.

Lesson three: follow up often. I started off with a basic e-mail service. It allowed me to collect e-mail addresses, send my messages, and interact with our fans. After six months, I knew I needed more. I needed a customer management system. After doing my research, I settled on a tool called Ontraport. Ontraport helped me collect phone numbers, personal details, and site engagement.

I sent interest-based e-mails according to website behavior and links clicked. This helped me follow up more often with parents and students. For our opt-in forms, I used a service called LeadPages. It has opt-in forms, lead boxes, and exit technology built in. Exit technology is put in place to capture a person's information if they try to leave the page. This powerful tool helps you capture more leads.

When I had everything in place, I experimented with different images, headlines, and forms. I achieved a 40-percent opt-in rate for our math calculators and a 25-percent-or-higher opt-in rate for our exam service. I celebrated this progress. In marketing and sales, micro-commitments are important. The more time people spend with you, the more they are invested. The more time they invest, the more time they are willing to invest. Investment leads to engagement, engagement leads to sales.

Because MathCelebrity is a tutoring service, we needed a free trial first. Free trials represent a powerful way to engage potential customers. I changed the process to allow users five free calculations. After they ran five calculations, the website prompted users to create an account. This account gave them a five-day free trial after which they had the option to buy. This way, it gave people plenty of time to test the service and see how it worked. This turned out to be an efficient process because I tracked users who logged in within twenty-four hours of signing up. Log-ins represent the first micro-commitment for a software based service. A free-trial user who logged in had the highest probability of becoming a paying customer.

Once I completed the account setup, I decided to start charging again. We stuck with the five-day-free-trial-to-paid route. We converted some of the free trials, but sales were still low. I continued for two months on this path. We picked up more than one hundred leads and free trials per day, yet we failed to convert them to customers. I couldn't understand why.

I spoke with an expert marketer and told him about my service. I read his newsletter for a few months and decided to call him. This phone call changed my entire mindset. This guy has sold more products and services in a month than most people do in a lifetime. We had a great talk where he confirmed what I long suspected. Our service is valuable. Our follow-up is strong. Our marketing is plentiful. However, there is one glaring problem: students do not have buying authority.

During this phone conversation, he told me one of his clients had a similar situation. This client tried to sell college recruitment materials to students. He struggled for months. He changed his strategy to market to the parents instead. When he did, he turned his business into a million-dollar-per-year machine.

Selling proved difficult unless I got Mommy and Daddy on the phone. Yet asking a teenager to make Mommy and Daddy grab their credit card is embarrassing. So I took another route. I decided to shut down all free trials and sign-ups and open the website up again. I changed my goal to focus on massive amounts of traffic instead. To generate revenue, I tried to sell extra services after people landed on the website.

After careful evaluation, I reviewed my future options. MathCelebrity needed a game plan shift to charge. A premium service required marketing to parents. Marketing directly to parents meant losing all our free traffic from Google. To do this, we needed a new website. This website needed a free trial, a sales funnel, and then a payment to create an account. I put this plan in my back pocket for a later date. I focused on building the existing website traffic before jumping to another website.

Marketing to schools seemed like an obvious choice. Yet it proved to be a poor strategy. Why? I left school marketing out of the story so far. But now it's time to tell you about our difficult history with schools.

SCHOOL STRUGGLES

Why don't you talk to teachers? Try visiting schools! If I had a nickel for every time somebody told me to visit schools, I would have my own jet. No matter what I tried, I struggled with schools (especially public schools).

From trying to set up meetings, to explaining what the product does, there seemed to be a constant tension with teachers. I gave free trials and built extra features, but it never seemed to earn me an endorsement with public schools. Important sidebar: this applies to the majority amount of time when MathCelebrity had free accounts.

An employee at a prominent high school in Illinois escorted me out of a lobby once. I walked into this school to give a quick introduction to the website. Now, this particular school openly advertised their plummeting math scores. When I arrived and stated my business, the person working the desk told me, "We do not need your help; we handle tutoring in-house." Are you sure? Because your press is telling the world a different story.

On one phone call, a teacher told me, "Showing students all the work is teaching them too much." You may think they told a joke, but they did not. How dare we lay out step-by-step instructions to help you learn quickly and easily!

Try setting appointments, you suggest? In eighteen months, I called one hundred public schools. I received a whopping two invites to present. Two measly schools! In one of those appointments, the meeting began with the math department head asking, "You aren't going to sell us anything today, right?" When I say the meeting began, I mean immediately after introducing myself.

After I gave my demo, the teachers told me we gave too much detail. One of them said, "This site is dangerous; it's too good. Thanks for coming, but we are focused on testing standards." At the time, my website covered 95 percent of the standards in Illinois. But, hey, who is counting?

When I spoke with a friend with ties to public schools in Illinois, he confirmed what I had long suspected: teachers treated this as a threat. They felt we encroached on their territory. They often asked, "Can't they use this to cheat?" I always responded the same: "I'm showing them exactly what a good tutor shows them—how to understand math problems." I am not, nor have I ever been, a fan of vague, halfhearted, go-figure-it-out-type teaching or instruction. I loathe it. Better to give too much information up front than too little. Students can filter out too much information. Filling in the blanks is much harder.

I almost forgot to mention one of the largest school districts in the country, Chicago Public Schools (CPS). During the summer of 2011, I called representatives in the education department. After a few follow-up calls, one of the people in the math department told me they loved the website. They had a few budget concerns, so they asked about the cost.

To ease their concerns, I offered the ENTIRE district sixty days of free trials. I included priority support requests for any new features they wanted. My website makes a difference in math, so I wanted to get it in the hands of every parent and student possible. Money turned into a secondary concern. I grew tired of hearing about poor grades and "lack" of solutions on the news. There is a solution out there—my website. To get my point across, I continued following up with my contact at CPS for a few weeks.

Then, out of nowhere, I started getting the voice-mail stonewall each time I called. No e-mails, no phone calls back, nothing. I offered to visit and give a personal demo and restated the "no cost" offer. All I heard were crickets. Before that, my contacts at CPS responded a fraction of the time. Eventually, I stopped calling them. Here is the best part: a few years later, the chief executive officer of CPS pleaded guilty to giving away contracts to educational companies of a friend. In essence, I wasted my time with CPS management. The fix was in from the start. No matter how good my website performed, it meant nothing in the end. Why? Because corruption, graft, and political entrenchment rule the roost in Chicago. If there is one thing I despise, it's having my time wasted. Here is what boggled my mind:

- Tell me you dislike the website.
- Tell me you have found a better competitor.
- Tell me you have limited time to look at my website.

But please, avoid wasting my time. To this day, CPS grades and performance are in the toilet. Parents are angry and budgets are a nightmare. If you are a parent in CPS, I want to help. I want to deal with you in person. I prefer dodging red tape and bureaucratic handcuffs.

I grow tired hearing about poor math scores and budget problems. The entrenchment at places like CPS hinders progress. What is in place right now is failing. Let's stop pretending there are no solutions. It's comical at best and pathetic at worst.

I have met some nice teachers who praised our service at conferences. Yet, when it's time to visit schools, the game changes. The pattern is always the same. A few teachers canceled appointments. Administrators skipped out of returning phone calls. Some of the teachers said the website revealed too much too fast. That's a verbatim quote. I grew tired of getting the runaround. Public schools have one of two choices: progress for students or protection of territory. They cannot have both.

I spoke with a few business owners in the digital education space. It appears other companies had the same problem. I had two choices: get a sales force to try harder with the public schools or just go around them. I decided on the latter. Why go headfirst into a brick wall when you can tiptoe around it?

There you have it. Those are the challenges of getting exposure in public schools. There seems to be an invisible wall built up in schools, so I avoid them for now and focus on search engines and e-mail marketing. Red tape and Common Core obsession are another problem. I know schools can benefit immensely from our service. Parental feedback confirms this. My plan is to avoid political entrenchment wars. My time is better spent on more promising avenues. I will dedicate my strategy to reaching students and parents through other channels.

Private schools were another matter. I want to be crystal clear about the dividing line between public and private schools. Private schools were courteous with meetings and open-minded about me demonstrating my website. A few private schools invited me to speak to the students as well. Principals, teachers, and faculty also distributed my postcards to parents. What a stark contrast from public schools!

If I could rewind time, I'd avoid public schools like the plague and focus on one group. I always have a blast working with them. Homeschoolers, I dedicate the next chapter to you.

CHAPTER 16

HOMESCHOOL LOVE

There is one group of people who I consistently bond with and have good conversations with. Hands down, the homeschool crowd takes the gold medal. From Great Homeschool Conventions to partnerships with websites like Homeschool.com, the conversation flows easily. It has been a great pleasure to meet everybody involved with those organizations. The best part is, I can have real conversations with real people.

The homeschool crowd is a joy to know. There is never any defensiveness, hidden agendas, or territory marking. Parents make their decision: either they like our product or they don't. Cut and dry is a good way to deal with people.

It is thrilling to watch homeschooling numbers climb year after year. The political environment and plummeting educational scores are to blame. It's no surprise homeschool parents voted with their feet by abandoning public schooling. They are taking matters into their own hands via homeschooling. One of the best quotes I heard came from a homeschooling mom. She said, "Instead of my son reading a history book in school, I brought my son to Mount Rushmore and showed him real history." This quote stuck with me for years because it puts things into perspective.

I think this is the primary reason MathCelebrity meshes well with homeschoolers. Parents can take their own lesson plan and use any section of our website as a supplement. There is no need for rigid lessons. Parents can direct overachievers to higher lessons such as calculus, statistics, and linear algebra. For parents who want their kid to learn the basics, they can start with basic math and pre-algebra. The feedback has been strong on this approach. Homeschool parents respect us for allowing unlimited subject access. They can go wherever they want on the website without subject restrictions.

The rise of digital education is another driver as well. Homeschooling parents have many choices online for education. Online schools, resources, and help forums only strengthen homeschool education.

Homeschoolers expressed another joy to me—not having to deal with Common Core. I admire the liberation of homeschool parents. They are free to teach their kids how to solve problems in any way they choose. Nobody rams Common Core down their throat. Homeschool parents are masters of their own domain.

I'm reminded of the quote in the movie *The Lion King*: "Everything the light touches is our kingdom." I feel a kinship with homeschoolers because they bucked the system by choosing their own path. I crave this freedom by running my own business. I respect freethinkers who carve their own path. Spend five minutes with a homeschool family, and you will see what I mean.

One thing impressed me so much about homeschool families, I wrote a blog about it. I published it on Homeschool.com, and it covered kids working on the family business at the conference. These kids worked in every facet of the business. They worked on sales, handled purchases, and spoke with attendees. Instead of being shackled in a test-taking-driven environment, they learned real-world skills. These business skills should be taught in school. How far ahead of the game in business do you think these kids will be when they reach age twenty-one? I would slay entire armies to have had that kind of knowledge early on.

If you work in education, you have to go to at least one homeschool conference and see this live. These families don't worry about test-taking requirements or Common Core. They simply teach, with the goal of providing knowledge best fit for their children.

While homeschoolers were shielded from Common Core, many students dealt with it head on. I heard some heart wrenching stories from parents. I'd like to share them with you, so let's enter the vortex known as Common Core State Standards.

CHAPTER 17

COMMON CORE: THE SCOURGE OF MATH

Halfway into the summer of 2014, a few parents messaged me on Facebook about Common Core. They hated it, and I mean Hate with a capital *H*. The requests from parents were the same: Please help with Common Core. We hate it. We don't understand it. Can you build something to help us?

Around this time, my wife told me a great story. Her friend's daughter received straight As in her math class. Up to this point, she had strong problem-solving skills. Unfortunately, the situation changed when her daughter started Common Core in the classroom. Her daughter came home depressed from school because Common Core confused her. Instead of allowing this girl to flourish, the school forced her to use Common Core. Because of this, her grades slipped. Even this girl's mother had no idea how to help with her homework. After my wife told me this story, I knew the time had come to build Common Core on the website. I found my call to arms.

The timing coincided with a national discussion on Common Core. The media centered on Common Core pains in the classroom. Public schools either abandoned or opted out of Common Core. Tensions boiled over on a daily basis. Imagine this scene: A student comes home from school in tears. In their hand is a math assignment littered with red marks from the teacher. These red marks signify incorrect answers. But here is the catch: the student answered the problem correctly. The student received a red mark because they avoided using Common Core to solve the problem.

Now I want you to pause and think about that. Picture this: You work at a job, and your boss tells you to finish a client assignment. If you complete the assignment, you get a million-dollar bonus. You work on the assignment for days and deliver it to the client. The client is thrilled; they sing your praises to everybody. Your boss calls you into his office. You strut down to his office to collect your bonus. But wait, the plot thickens. Your boss delivers the bad news. Congratulations, you completed the job, but you failed to use my silly standard. As a punishment, you lose your million-dollar bonus.

Even though you have a satisfied client, you lose your bonus on a technicality. Welcome to Common Core logic: where correct answers are decimated for political reasons. No matter whom we talked to, we felt the frustration and rage just oozing out.

Parents had a good point: if my kid gets the right answer, who cares what method they used to get there? Parents and schools butted heads over this issue. The search for a workable solution continued.

No matter whom we spoke to about math, Common Core always came up in conversations—always. I think we should rewrite the famous saying about death and taxes. There are three certainties in this world: Death, Taxes, and Common Core frustrations. It remains one of the most polarizing subjects you will ever hear people talk about. I believe Common Core ranks up there after politics and religion on the conversational polarity scale. The debate rages nationally as I write this.

The *New York Times* published a 2015 article entitled "Nationwide Test Shows Dip in Students Math Abilities." The article discussed the drop in America's math performance for the first time since 1990. The suspected culprit? Common Core questions on the exam administered by the National Assessment of Educational Progress. The article states, "The largest score drops on the fourth-grade math exams were on questions related to those topics."

Parents asked where I stood on the matter. I'll say now what I said then: I'm not a fan of Common Core, and I don't think it should be forced on students. But if it works for one kid in a classroom, let that kid use it. I'm a fan of the traditional method, which is how our parents learned. At the end of the day, nobody should limit a student to one problem solving method. As long as they get the right answer, nothing else matters. Who cares what method they used to get there?

Now allow me to build my case. This is another point I vehemently disagree with when it comes to high school and college. At the end of the day, students are going into the workforce. For projects, clients and bosses care about the end result. How you arrive at the finished product does not matter.

Let's do a little experiment: Put two experts into a room on any subject. They both produce an outstanding end product, but they use completely different methods. Why are we punishing one person over another? Forcing everybody to use one process is robotic and destructive to creativity. This is my biggest disagreement with forced Common Core in the classroom. It handcuffs problem solving skills.

Close your eyes and think for a moment: What is the end game of math? Problem solving. Even if your chosen field has nothing to do with math, problem-solving skills apply to all careers. Math, like life, is a problem-solving game. So long as the problem is solved, who cares what method a person used?

With the torrent of emotions running wild, I made a choice. I began building as many Common Core calculators as possible. Yet I needed to tread carefully. Our loyal fans who had no obligation to use Common Core may get upset if the website displayed it.

I decided to expand our choice-based calculators. If somebody entered a problem with a traditional *or* Common Core method, I gave them a choice. I call this choose your method. I spoke with a few fans who used the traditional method and this idea resonated well.

New Common Core calculators presented a challenge. It took me three days to wrap my head around the Common Core logic. Concepts like estimating fractions tested my patience.

The problem seemed to be my bias toward traditional math. To remedy the problem, I focused on removing my bias. I took the first difficult step: unlearning traditional math. The quote from the movie *Training Day* speaks volumes here. Denzel Washington is a cop leading an elite police unit. He is driving around in his car with his newest trainee, Ethan Hawke. Denzel tells him, "Unlearn everything you learned at the academy. That will get you killed out here!" This "academy" symbolized traditional math while Denzel's team represented Common Core. Common Core pushed me into uncharted waters. I had to learn all over again.

First, I read a few books on Common Core to get myself acquainted. I crafted a simple plan: read, comprehend, and start mentally building the algorithms needed. These algorithms worked exactly like any other calculator on the website. I take all parts of a problem and build an abstract method to solve those problems. Once I read and comprehended the lesson, I built a simple spreadsheet with input values and an output answer. After completing the spreadsheet, I wrote the code to build the website calculator. I wanted to deliver a methodical and easy-to-follow approach.

People were desperate for a clear, structured approach to solving Common Core. Other sites provided confusing documentation. We had a chance to step up our game and simplify the difficult.

I started building Lattice Math multiplication. Parents mentioned this lesson contributed to Common Core pains. Next, I built estimation of fractions followed by number bonds. Once I built these three lessons, I notified our fans. Shortly after this, the Common Core tickets began to pour in. Imagine a large dam bursting and water gushing out. That is how the ticket list looked after I built the initial three Common Core calculators. After completing parental requests, I let the ticket report guide me going forward.

Because true-false equations dominated our ticket report, I made this my first priority. The ticket report filled up with problems such as 10 + 4 = 12 + 3. A few families I spoke with hated these types of problems. Parents said things such as the following:

- Why do kids need to know this?
- What exactly is this problem asking?
- If there is no example problem, how are parents supposed to know if their kid did it right?
- I felt so stupid. I messaged the teacher because we struggled to figure it out.
- 14 does not equal 15!

I needed to build the calculator with clear math work to determine why an equation was true or false. I also included a practice-problem generator and flashcards so students had plenty of examples.

I spent the next six months completing Common Core tickets and going through books to build everything possible. Our calculator list grew and our functionality improved. One day in the fall, the daily Common Core tickets started decreasing. This indicated to me we finally built a solid Common Core offering.

As our Common Core offering grew, our mentions also grew. Our fans credit MathCelebrity as the go-to calculator to handle the scourge known as Common Core. Our reliability, service, and speed pay dividends to this day.

While speaking with parents about Common Core, I learned a valuable lesson. While automated tutoring is great, sometimes students need personal help on a homework assignment. They require a human element, above and beyond a computer. What a coincidence: I know just the guy to help.

CHAPTER 18

PERSONAL HOMEWORK HELP

As the website grew in popularity, a few students asked for personal help on homework and quizzes. While the website covered a vast range of subjects, three personal homework requests topped the list:

1. Statistics
2. Algebra
3. Programming

College students craved statistics help. Between confidence intervals, hypothesis tests, and distributions, some students were at wits end. Tight deadlines aggravated the situation for students even more. I worked with a few students to set up a pricing structure and a sales page. I called it the homework coach program. I structured the homework coaching program for one-on-one help. The sessions included a brainstorm approach to problem solving. This approach involved me talking out the steps and questions I ask while solving a problem.

The challenge with confidence intervals and hypothesis tests is deciding which approach to use. Sometimes, the process is complex. You must choose a distribution, set up a test, and state your conclusion. This intense process created a prime opportunity to help students understand the steps involved. I priced the service to reflect the quality delivered.

Some of the students disliked the high price, but I had no regrets. The time savings and knowledge gain are worth it. For the students I helped, the feedback and results justified the price. Large assignments and tight deadlines are doable, but it is a high-value service, hence a high-value price. If somebody wants to price shop, call Walmart. I deliver value and I don't haggle on price. Those who purchased received high-value homework help.

These one-on-one sessions proved to be a valuable website content generator. It felt good returning to my personal tutoring roots. I learned more homework and exam sticking points from students. After the homework session ended, I added notes, tips, and tricks into the relevant calculators. Homework sessions improved the calculators, and the calculators improved the homework sessions. This resulted in gradual and consistent improvements—the *kaizen* way.

Algebra homework sessions involved equation solving, word problems involving equations, and quadratics. People shared their struggles with me during these homework sessions. The concerns they had matched up exactly with the survey results. Students understood bits and pieces of concepts, but they had two distinct difficulties: starting the problem and linking the concepts together. While math had unique difficulties, other homework subjects brought new challenges. My next challenge: getting students into the mode to code.

Programming homework sessions involved C-languages, MATLAB, and Excel/Visual Basic spreadsheets. Students told me they struggled with how to get started while programming. They needed help with syntax, functions, and structure. I used the same process with programming as I did with math. I delivered a step-by-step process leading to a desired result. I detail the thought process below:

1. Where do I start?
2. What variables do I need to define?
3. How do I repeat steps?
4. How do I use a piece of code over and over again?
5. What is the end result?

I added comments in the code and step-by-step instructions for these situations. When a student came across one of those five issues, they had a checklist of steps to complete their code.

There seemed to be a nice intersection of math and programming in the assignments I helped with. Some example assignments are below:

• Write a function to find the first twenty even numbers.
• Write code to calculate the value of a mortgage payment.
• Write a function to determine the first fifty Fibonacci numbers.

As an interesting coincidence, these three homework assignments are also calculators on MathCelebrity. I showed my students how programming uses math concepts. I detailed this through various programming examples.

While sporadic homework coaching helps, certain students requested additional help. These discussions led to the creation of the Homework Mastermind.

CHAPTER 19

HOMEWORK MASTERMIND

I discovered a small percentage of students I tutored needed more frequent homework help. These students wanted access to me all the time when they needed help. This led to the creation of a homework mastermind. The mastermind allowed a small group of students to contact me directly whenever they had homework issues. I also developed the mastermind to bring a group of students together to watch problems being solved in real time. I believe strength builds in numbers.

Masterminds have a proven history of increasing knowledge and stimulating creativity. They go back to the early 1900s. Masterminds foster a sense of unity. This unity creates intense focus and shared vision to solve problems and generate ideas. Napoleon Hill discussed this in his famous book *Think and Grow Rich*. The true power of masterminds comes from exponential growth. You learn faster and grow more in a small group of like-minded, motivated individuals. The power of the group amplifies the power of the individual.

I structured this mastermind as a high-end homework coaching group. I limited membership to a small number. I charged a high fee so students who needed extra help took it seriously. They appreciated the exclusivity. The mastermind membership granted priority scheduling on my calendar. This exclusive access instilled confidence when students needed it most. The mastermind put me one phone call or text message away when students required help.

I enjoy reading, and many of the books I read talk about the benefits of having a coach. It boils down to accountability. Somebody is there to make sure you stay on track and achieve your goals. An exclusive group ensures you get the attention you deserve. Weekly reporting forces you to improve performance and keep your mind sharp.

Exclusivity generates powerful ideas. These mastermind sessions gave me insight into the homework troubles students face each day. I discovered powerful ideas from these sessions. One of these ideas turned into an unexpected new line of business for MathCelebrity.

AN UNEXPECTED NEW LINE OF BUSINESS

The next day job I took involved data and analytics. At this job, I began working with Google Analytics. Google Analytics tracks activity and visitor engagement on your website. With my momentum focused on marketing and sales, this fit perfectly.

One of the key points you learn in sales and marketing is what gets tracked gets improved. Dan Kennedy hammers this point home in his book *No B.S. Ruthless Management of People and Profits*. This goes for any endeavor: weight loss, sports, making money, website traffic, and so on. In business, I live by this motto. I wanted to improve my website, so I decided to get certified in the Google Analytics exam. One of the parents I met through the homework mastermind mentioned the analytics exam as well. She needed to get certified for her day job.

After studying for a few hours, I felt comfortable with the material. I put together study notes from what I learned. The next morning, I took the exam and passed on the first try.

Later on at dinner, the thought crossed my head: how about adding the questions and study guide to a page on the website? The study process takes time just like math homework. The rewards for passing are good, whether it is promotions, bragging rights, or a badge of expertise. While it's not math, it is education, a core premise of MathCelebrity. What's the worst that happens? Nobody visits the page? At the least, we get new traffic. In addition, there may be potential cross-promotions for math and exams. Who knows?

It reminded me of the palliative care study done by a nurse. She tallied up the data and listed the top five regrets people had. One of the five related to taking more chances and trying more things. I've never forgotten this study. Something told me this exam had serious profit potential.

I submitted the Google Analytics exam page to Google's search index. Next, I created a free study guide detailing the basics of Google Analytics software. This study guide contained three correct answers on the exam. To get this free study guide, I built a sign-up form on the website. If you signed up for the free study guide, you had the option to buy all the exam answers immediately or at a later date.

I decided to charge $24.99 for all the exam answers. Here is how it worked: A user came to our website and saw all the possible questions and answer choices on the exam. We had one catch—we concealed the correct answer until you purchased. As a bonus for customers, we gave a link to the answer source. Imagine being able to visit a page on the web and find how we got the correct answer. This confirmation link became a one-click learning session for the user.

The exam service created one of those finer asymmetrical bets in life: a limited downside with plenty of upside. If it worked, we had the ability to build a new fan base. The exam service had potential to become an income-producing asset and lead generator. It allowed MathCelebrity to expand the business. Keeping current with the answers added a challenge. To do so, I built a three-month automated reminder for myself to get new questions and answers.

The next morning after building the exam page, I woke up to feed the dogs. While they were eating, I checked e-mail with my eyes half-asleep. I had to pinch myself to see if I hallucinated. Five people signed up for the free study guide. Three people purchased the exam answers immediately.

This happened only twenty-four hours after I set up the Google Analytics exam page and submitted it to Google. I checked the Google Webmaster Tools report. This report monitors search activity related to your website. What I found blew my mind. Six of the questions on the exam already ranked on the first page of Google. Two of the six questions ranked in position number three.

Best of all, I did not spend a dime to get a first page placement on Google. Instead, I have our page setup and domain authority to thank. You see, I mistakenly assumed people searched for the phrase "Google Analytics exam." Instead, they searched for the entire question on Google. I set up my pages to have the question as the page title. When people searched for the question, my website popped up. The description text on Google showed the exam question. From there, the free traffic poured in.

This revelation clarified how users search on Google. People searched for the question because they were in the middle of taking the exam. They needed help as the clock ticked away. Because I understood the user thought process, I had to update the website. I changed the Buy button text on MathCelebrity to address urgency. The button read, "Buy all answers and get CERTIFIED IMMEDIATELY." This wording gave users confidence in our delivery speed if they purchased the exam answers.

One of our customers who purchased the exam certification answers notified me about missing questions. Thankfully, there were only two questions missing. He passed the exam with ease using our website. During his exam, he had the presence of mind to take screen shots of the missing questions. I added those questions with correct answers to our exam page. I resubmitted the updates to Google and waited.

I had my ego crushed before, so I wanted to wait a few days to check if this momentum continued. The next day, more people purchased, and I received two e-mail support tickets:

1. Google Analytics exam is great. Do you also have Google AdWords?
2. What about a Bing exam?

What a rush! I've established demand is real, and people need exam help. Because our fans wanted more, I had to deliver. It took me another week to get the Google AdWords Fundamentals exam set up. I passed the exam, created the study guide, and added the questions to the website. I set this up exactly like the first exam.

I submitted the AdWords exam page to Google. The same thing happened with this new exam as did with Google Analytics. Welcome to déjà vu! We had more exam answer purchases and new sign-ups for the free study guide. MathCelebrity now had two exams under its belt. From here, I had to add more exam certifications. To promote this new exam service, I added another section on MathCelebrity dedicated to exam certifications. I wanted to make things crystal clear to people who came for math only versus exams.

From this point, I had it made. I compared this to an oil well. Once you drill a successful spot in the ground, you milk it for all it is worth. Search engine marketing exams became our oil well. So I added Bing next, DoubleClick, and the requests kept pouring in. Just like with math, our fans told us what they wanted, and we built it.

I decided to offer an exam bundle as an upsell. You pay one monthly recurring fee and get access to all the exam answers. We had a nice conversion rate on this product. As we added more exams, the exam bundle increased in value.

I also found a solution to keep current with exam questions. While I took the test myself every few months, our fans provided ongoing support. When they took exams, they alerted us about questions we lacked. This way, I had a small army of eyeballs who supplied updated questions. I spent months building up trust equity by giving away free information. Our fans repaid my loyalty back in kind. What a beautiful setup to have.

The exam certification feedback provided motivation to push forward. Some customers got raises and bonuses from passing exams. Students performed better in marketing classes. They learned more about marketing, search engine optimization, and analytics. Because we showed the answer source, our service acted as a real-time study guide. I loved the feedback, but I needed more information.

I started surveying our exam certification users. Two of the questions I asked them on the survey were very telling:

1. Why are you getting certified (job, school, bragging rights, etc.)?
2. What is your biggest fear/concern about taking the exam?

For the first question, "For my job" ranked number one at 65 percent. "School assignments" ranked number two at 30 percent. I learned how marketing students in college needed to pass an exam certification. As usual, another eye-opener only a survey provided. For question two, the number one fear/concern for exam taking—running out of time.

This pleased me because MathCelebrity specializes in speed and efficiency. Our mission is to help you learn as fast as possible. To capitalize on speed, I built an instant search box on the exam answers page. When a user began typing, the list of questions automatically filtered. The exam service provided a quick way for the user to get what they wanted. Building this feature saved a few seconds per question, which is a significant time shave on an exam.

As the exam service grew, I wondered whether having both lines of business on one website created a conflict. From time to time, I considered splitting off the exam certification section into a separate website. Right now, I'm keeping them together. It's profitable, reproducible, and best of all, it's a passive income source. We educate people with speed as a primary benefit. What's not to love?

Besides, keeping the exams on MathCelebrity uses our domain authority with search engines. This way, I don't have to go find free traffic again with a new website. When I add a new exam page, we get the Google boost from MathCelebrity. Speaking of Google boosts, let me tell you how we get hundreds of thousands of people to our website for free.

CHAPTER 21

HOW TO GET TWO HUNDRED THOUSAND MONTHLY VISITORS: MY FREE TRAFFIC FORMULA

I often get asked how MathCelebrity gets so much free traffic. Over the last few years, we have enjoyed serious traffic on Google without spending a dime. I give all of these secrets away in a free-traffic-secrets course. There are a few successes I can share in this book.

When people visit our website, they often return. So once we get them to our website, we benefit from a high return probability. We also have people linking to us on other websites. When they like our website, they tell others via word of mouth. However, word of mouth represents six percent of our monthly traffic. The lion's share of MathCelebrity traffic, 94 percent, comes from Google searches. Correction: make that FREE Google searches.

I've tinkered with the website layout periodically, but one strategy delivers consistent website traffic. I use it ruthlessly and it never fails. I present to you free-traffic secret #1:

$$P = T = M = H_1 = B_1$$

- P is one key phrase summarizing your entire page.
- T is the page title.
- M is the meta-tags.
- H_1 is the main header text.
- B_1 is text somewhere in the first paragraph of the body on the web page.

The first step is determining the P. I like to keep these phrases short. One exception of the short-phrase rule is a question for exam certifications or a math word problem. Some example phrases are

- Interval notation calculator
- Synthetic division calculator
- Or a Google Analytics question like, "Which of the following would you use to set up a custom alert?"

First, pick one phrase describing your page. Make sure your entire page is built around this phrase. If you have related content, set up a link pointing to this content.

On MathCelebrity, the Literal Equation Calculator references interval notation. It makes sense to set up a link to interval notation. As you build more content, your website turns into a hub and spoke model. Picture a wheel. The hub is the main section. Multiple spokes connect to the hub. When you build more related content, you build more spokes.

Another great analogy for this is a spider web, where your company is at the center of the web. The more relevant content a person finds, the longer they stay. The longer they stay, the more they trust you. The more they trust you, the more they consider doing business with you.

What about generating new content ideas for your fans? I don't like to sit back and hope people like a few pages on my website. User surveys provide ideas. However, surveys require faith in responses. I have a better way: your fans tell you what they want. For this, you need a system. What kind of system, you ask?

Free-traffic secret #2: Track every internal search on your site. A search is a digital vote. Make sure you track and analyze all votes. Set up a reporting feature to handle website searches. View the report the following morning. You can use Google Analytics or log searches on your site using a custom method. Why is this beneficial? Because the user gives you ideas for content—for free.

On MathCelebrity, people search for math problems and math terms. They also search for science terms, programming terms, and exam certification questions. When we don't have what they want, I build it within twenty-four hours because there is demand.

Imagine if your customers approached you and told you what they want with lucid detail. You build this, and they pay you for it. They spread your message around the world. Your content expands, your search rank improves, rinse, and repeat. This is the holy grail of website growth. Your fans hand you the keys to the kingdom; all you have to do is unlock the door. This strategy has proved to be lucrative for our website, and I use it obsessively to this day.

Using the tracked search approach, I examine our ticketing system for missed searches. A missed search means a user ran a search our website failed to process. We simply lack the material they want at the time they run the search. This capture box gives them the option to add their e-mail address for notifications. When I build the new page or feature related to their search, the system e-mails them. This encourages the user to be more invested in the website. It also demonstrates your rapid response time. Now the user gets real-time updates about fixes to the website you made based on their suggestion(s).

There is another benefit to this notification system—trust. Trust in your website and trust in your ability to deliver quickly. I remember one of our fans giving us five calculator ideas in the course of two weeks. It all started with the response time to the first ticket she logged. This fan loved our quick response time. The automated e-mail started a conversation with this fan, which led to more ideas. Not only did MathCelebrity get five new ideas from one person, but we also found a new ambassador. After building the five calculators, this fan told her friends. Thank you for the free publicity. You see? Internal search monitoring and delivery speed made this possible. This level of service inspires people to become ambassadors for your brand.

When you develop your brand, treat your ambassadors like royalty. If you find somebody who enjoys your site and understands how it works, get ideas from them. If they are good ideas, build them and include your new ambassador. I say this because one ambassador can spread your message far and wide. People value feedback from their peers much higher than any self-promotional material you produce. Look at Amazon ratings and look at Yelp. The peer review system dominates the digital landscape. Ambassadors also have the trust of friends and peers to provide you with more ideas and feedback. It's a perfect circle, and one you should embrace. Listen to your users' concerns. Listening and responding builds trust. Without trust, you are just another page on the Internet.

Free-traffic secret #3: Watch how people search for things on your website using their idiosyncratic phrases. People have many ways to say the same thing. Identifying language and speech patterns provides insight into your customers. Phrase structure and slang give you ideas on how to structure your page. Spelling errors give you more ways to link to content. This way, you avoid missed searches on your website. Plugging these holes on your website earns you more traffic and keeps people on your site longer. Some misspelling examples from MathCelebrity are below:

inturval $\Rightarrow$ interval

millograms $\Rightarrow$ milligrams

By correcting common spelling errors, visitors stay on my website for an average of three minutes. Users visit five pages on average when they find what they want on the first try. Imagine what this does for your website. Let the user guide you. Why fight it?

To perfect this process, I built an autocomplete search box based on website concepts. As the user started typing, they were given a list of suggestions. When a user has to type less, three great things happen:

1. Users find what they want faster.
2. Users type fewer spelling errors.
3. Users know they are in the right place.

Your users are your market. Tailor your website to user interests. I never argue with the market. In fact, there are two things I never argue with:

1. Mother Nature
2. The market

When Mother Nature throws her full weight behind a weather phenomenon, I don't argue. I stand clear until her fury ceases. The same goes with the market. The market is going to go where it wants to go. Instead of fighting gravity, I watch what the user searches for. Tracking internal search requests gives me content ideas and expands our reach. Staying on top of this report is a big reason for our organic search-rank success.

I learned to treat searches like votes. When a user searches for anything on your website, it's a digital vote for a concept. Make sure each vote counts.

While I have made some good decisions on how to get free traffic, I made poor ones as well. I cannot share the good without sharing the bad. One decision in particular almost destroyed all the free traffic we received.

CHAPTER 22

IGNORING INNER VOICES LEADS TO STUPID CHOICES

No matter how disciplined you try to be, lazy and arrogant moments rear their ugly head from time to time. Two of them come to mind; they are impossible to forget. In October 2013, Google turned my free-traffic lights out. Its algorithm update penalized our website, to the tune of a 40-percent traffic drop. Up through October 2013, my traffic exploded from strong rankings on long-tail key words.

Then I got greedy. I thought, *Why not get more links but faster? The more links, the better, right?* I bought some backlinks from a Fiverr gig. If you don't know Fiverr, it's an online marketplace where you pay five dollars for random tasks, which are called "gigs." The backlink gig I bought promised to get my website featured on various blogs—five thousand links in seventy-two hours. Yup, sounded good to me. I told myself we had a valuable website. The more people who knew, the faster we succeeded. How about taking the escalator to success instead of the stairs?

Man, what a dumb move. Google released another algorithm update on its search engine. This update targeted links built on spam-based websites. Can you guess the quality level of the Fiverr website backlinks? You guessed it—absolute garbage. Traffic plummeted, sliced in half overnight.

I consider this the worst mistake I ever made with this website. I tell this story to warn others. If you are working with backlink services, please be careful. You are much better off building organic links. Remember, links are votes, so don't rig the voting machine.

It took me a year of work and a king's ransom to undo the damage I caused. Over the next year, I cleaned up most of the garbage backlinks. I acquired new legitimate backlinks with blog postings and media placements. I had a radio interview with Mark Imperial at Remarkable Radio, which went well. This produced some nice media exposure. We hosted a one-thousand-dollar Amazon gift card giveaway. This gave our brand a boost, but our traffic recovery continued moving slowly. After fixing the backlink problem, I ran into another problem: mental fatigue.

After six years of working on the website, I grew tired of responding to tickets. Between working, running the website, and learning about sales and marketing, I ran out of gas. My motivation sunk to all-time lows. I went through a two-month stretch where I had zero motivation to work on tickets. Working on search engine tickets felt like a chore. I scratched and clawed until one day, I gave up trying to keep up.

Algebraic expressions kept popping up on the nightly ticket report. I built this calculator two years before, yet more algebraic expressions continued to appear. This calculator annoyed me to no end. Think of it as a perpetual cold; it never goes away. Speaking of colds, here are a few examples of lingering algebraic expression tickets:

- Two less than 5 times a number
- Six more than 3 times a number
- Nine more than $3y$
- Thrice a number
- The quotient of 5 and $2x$
- The product of 8 and $3x$

Night after night, two or three of these types of problems slipped through and wound up on the dreaded ticket report. During my lazy months, I saved them on a Google document. The problem remained—I failed to act on the tickets. Just like the song in the movie *Top Gun*, I'd lost that loving feeling. Keeping up with this madness proved to be overwhelming. So I mentally checked out. I relaxed for the first half of the summer.

I needed this break, no doubt about it. In late June, my wife had dinner plans with friends, so I stayed home to chill. I fixed a Rusty Nail, and then turned on my iPod on random mode. Ten minutes into random mode, one of the house music tracks came on. This song came from the playlist I listened to during the first three years of building the website. Between the buzz from the drink and the Chicago house music, I felt bulletproof.

I opened my laptop and scrolled through the ticket report. Those algebraic expression tickets glared at me. Maybe it was the scotch or the house music: I suddenly became interested in solving this problem. After six weeks of dodging updates like Neo dodging bullets in *The Matrix*, the time arrived to fix these problems. I copied eight of the algebraic expressions from the ticket report to an Excel spreadsheet. I looked for a pattern or similarity to appear.

After thirty minutes of staring at the spreadsheet, the answer appeared:

- Numbers written in text should be converted to a digit representation. For example, three becomes 3.
- Categorize each operation and the respective terms.
- Addition is written as: and, sum of, plus.
- Subtraction is written as: difference, less, minus, subtract.
- Multiplication is written as: times, product, multiplied
- Division is written as: quotient, divide, over.
- Each algebraic expression may have an extra assignment portion: equals, greater than, less than, greater than or equal to, less than or equal to.

For the next two weeks, I worked on this lesson diligently. I built shortcuts to identify patterns and blocks of code to test each pattern. After the two weeks of upgrades, the daily algebraic expression tickets decreased 50 percent.

Once I got algebraic expressions under control, I tuned up another calculator: Order of Operations. Our fans called it PEDMAS, which stands for the following:

- Parentheses
- Exponents
- Division
- Multiplication
- Addition
- Subtraction

My wife brought this calculator to my attention after seeing people ask for help on Facebook. While I wanted to relax, my wife put the pressure on. She felt compelled to post this calculator on Facebook as the solution to PEDMAS problems. I admired her dedication to promote the website. There have been countless times where she finds somebody in need of math help on Facebook. She uses the MathCelebrity calculators to help people find a solution to their problem.

It seems everybody had an opinion on PEDMAS. Teachers and math tutors e-mailed their personal interpretation of this lesson. These e-mails often led to entertaining arguments and colorful language. Two months of discussion led to a solution.

I embraced simplicity as the solution to my problem. I took one letter at a time, performed the operation, and moved on to the next step. Solving each letter in PEDMAS individually fixed the problem. This method described clear instructions on the calculator.

I rolled out updates over the next two months. I e-mailed fans who wanted updates on changes for their personal PEDMAS problems. Each upgrade to the Order of Operations Calculator pleased our fans.

While fans were happy with the calculator updates, MathCelebrity lacked something. People were polite with us about it, yet the problem remained. I received hints about it over the years, but it finally came to light one day while reading. I stumbled across an article about user experience. My daily reading sessions uncovered our weakness.

CHAPTER 23

USER TESTING: THE COLD PUNCH OF REALITY

Continuing my marketing and sales education, I found a website called UserTesting.com. It provides live user feedback on your website or application. You write up a list of tasks for a person to perform on your website while recording the session. One concept they stress is user experience. When a user lands on a site which is confusing or poorly designed, they are more apt to leave. This is why design is crucial to retaining website visitors.

I avoided design feedback for the first five years. I had no surveys and no chat software. I assumed people liked the website and found it easy to use. One article I read on UserTesting put my theory to rest. It described a business owner in the same situation as I was. He purchased the UserTesting service and learned a treasure trove of information. The true value came from the user's reaction during the test. You heard their voice and watched as they hit roadblocks trying to find things on the website.

I admit, fear prevented me from doing this before. As a business owner, arrogance and pride creep in and prevent you from confronting unpleasant truths. Because sales were low and there were still many unknowns, the time came to pull the trigger and buy a test on UserTesting. I had to learn more about our users and their experience on our website.

The UserTesting interface gives you suggestions and popular questions to ask the tester. A fifteen-minute test costs $50. I purchased a testing session and logged into the system. I scanned the site and came up with the following tests:

1. Have a user visit the home page. Next, have the user sign up for the e-mail list.
2. During this test, have the user try to identify what the website does.
3. Next, have them identify their initial thoughts on professionalism of the site.
4. Run a calculation and give their feedback.
5. Visit our sales page and give their feedback.
6. Go to the checkout form and give their feedback.

I set up my first test and waited for the results. Within one hour, UserTesting sent a completion notification. When I opened the notification email, I felt a spark of fear. I put off a professional design for too long, and my mistake came back to haunt me. As the old saying goes, the tape never lies. I clicked Play, and the video started.

When the user visited the home page, they griped about three things:

1. The user groaned about our menu and how we had different sections for math and exams. The user had trouble figuring out what our website did based on this menu.
2. The user needed a clear message of what we do on the home page.
3. The user said the home page looked like "website from the 1980s."

Next, the user visited a calculator. I received more feedback on the lack of professionalism. We scored high marks on functionality, content, and concept. This came as no surprise even though hearing it on a live test felt good.

After visiting a calculator, the user navigated to the math tutoring sales page. We got hammered again on the professionalism and the clarity of what we sold. One of the questions I asked was, "Do you trust this website with your credit card information?" The user gave a resounding "no." The user criticized the design, colors, and layout.

I grabbed a shovel to pick my ego off the floor as the user finished the test. Because I hate design, I needed to hire a professional designer. We needed a website design focused on user experience, message clarity, and professionalism. I procrastinated for too long. Our new design must reflect the value MathCelebrity provided. MathCelebrity had a Ferrari engine with a station wagon paint job. We had to do better.

We needed a mobile responsive design as well, sized for all screens. Google provided additional motivation to design a mobile-friendly website. Their latest update boosted the search rank for mobile-responsive websites. For many years, I focused on desktop design only and ignored phones and tablets. That strategy turned into a loser; it had to be fixed.

We lost sales, rank, and respect because of our design. Thousands of potential buyers abandoned us. We had a great product, but our look and trust factor killed us. This turned users off, reducing time on the website, product purchases, and faith in the brand.

After deciding to redesign the website, I had to find somebody to help me. Through my sales and marketing training, I had the perfect person in mind.

REDESIGN

I joined Matt Ackerson's subscriber list six months before the UserTesting results. He runs Petovera, which specializes in sales funnels and conversion-based website building. Prior to working with Matt, I discussed my problem with other companies. Unfortunately, these consultants were clueless. They kept harping on getting more traffic. I get two hundred thousand unique visitors per month. I don't have a traffic problem. I have a CONVERSION problem.

I set up a call with Matt and explained the UserTesting results. After speaking with Matt for ten minutes, I knew he had the expertise to help us. He understood my product, user complaints, and the work required to resolve those complaints. We settled on a quote, and his team got to work.

Within a month, his team completed the redesign. The finished product looked outstanding. The website had a cleaner navigation menu, highlighted search box, and mobile-responsive design. The new sales page looked elegant with a clear message describing the benefits of buying our services.

Within three weeks, people commented how clean the website looked. They noted how easy they found what they were looking for. Google Analytics showed increased engagement, decreased bounce rate, and lead growth. The key traffic and lead-generation metrics increased across the board.

I worked on improving the checkout form next. Each product and service needed a vibrant image, clear description, and bullet-pointed benefits. I incorrectly assumed our fans remembered the benefits from the sales page. Restating these benefits on the checkout form ensured clarity and continuity of our message.

I signed up again at UserTesting.com and bought another test. This time, we received strong professional scores. The user found what they needed easily. The video session contained no confusion on where to find links.

More fans gave feedback the next month. Our fans noted the clean layout, look, and clarity of the website. It felt great to get high marks on professionalism. Our fans gave us positive reviews on the redesign. The website looked far better on mobile devices. The investment in redesign helped our brand.

After the redesign project completed, I focused more on a high-level overview of math. I chose three topics to learn more about:

1. How does math fit into society?
2. What challenges do students face?
3. How to improve the math experience?

I settled on a survey software tool to get us the research we needed. The research I gathered from this survey gave clarity to our mission.

THE STATE OF MATH:
PUTTING THE *M*
IN STEM

In April 2015, I added an online survey on the website using a tool called Qualaroo. It allows a website owner to build smart survey questions, where the next question depends on the answer to the prior question. The tool helps a website owner learn about audience pain points. I collected anonymous surveys from more than five thousand students and parents. The survey results fly in the face of traditional assumptions about difficulties in math.

One of the survey questions I ask is: "What is your biggest challenge in math?" To begin, I made this field a text box where the student or parent typed in anything they wanted. After two months of gathering answer details, I classified seven types of student math challenges:

1. Difficulty understanding the material
2. Takes too long
3. Boring: I am bored to tears

4. No challenge: it is easy
5. I'm decent in math; I want to check my answer for reassurance
6. I just need help starting a problem
7. Common Core

I took these seven types and changed the "biggest challenge" survey question to a multiple-choice answer format. The results after six months were fascinating. Number one: difficulty understanding the material. This clocked in at 50 percent of survey responses. Number two: I'm decent in math; I want to check my answer for reassurance. This accounted for 25 percent of the responses. Number three: I just need help starting a problem. This came in at 20 percent of responses.

Currently, there is a myth running rampant in educational circles. The myth states any student struggling in math struggles due to intimidation. Let this survey be the myth corrector. A chunk of students needed help at the start of a problem. My personal homework sessions confirm this. Give students a nudge and a hint at the beginning to get the ball rolling. Then let them surprise you with their ingenuity to finish the problem.

If you think about it, this makes perfect sense. How many projects have you been on where you needed a simple push to get started? Why do you think the best training programs have example documentation? They not only show you how, but they also show you why. Has there ever been a time where one little hint or spark of information moved you to act? The survey results reflected real-life projects.

The number two survey result sparked my interest as well. Students wanted confirmation for their calculation. Students learned more after comparing their solution method versus MathCelebrity. This also makes sense given how human nature works. We look to our peers for approval and verification. Peer approval is as old as the hills.

The last survey result should turn educators' heads as well. The following two survey results can be combined into one: boredom and math being easy. Gamification provides one potential solution to boredom. Gamification is white-hot right now for achievement-based websites and video games. This is because medals and badges get people excited to reach a goal. It keeps them engaged and motivated. Think about a simple game like one-on-one basketball. It's about more than winning; it's also about bragging rights. Competition provides drive, and drive creates improvement.

The information taken from this survey will serve math tutors as well. This powerful marketing tool gives tutors insight into their clients. Forget paying $10,000 or more for market research; here it is on a platter. And the best part? Students and parents answered questions WHILE they worked on math problems. If they took this survey on a weekend while relaxing, their mindset would have differed. I gave the survey during the homework sessions to capture their mindset in real time.

General market research surveys take place in some air-conditioned sanitized environment. These generic surveys are only filled out for the cash prize at the end. MathCelebrity surveys worked differently. I gave no gifts or bribes in exchange for the unfiltered feedback. You won't find any Ivory Tower theoretical fluff here. These survey results come straight from the mathematical trenches.

Math departments should plaster these survey results on every wall in their school. Math tutors should use these survey results while marketing to parents and students. Teachers should grab megaphones and shout these results to everybody who will listen. The survey results are another example of data telling you a different story than what traditional media rams down your throat.

We have talked about the students; now let's focus on the parents. Parents lie awake at night wondering two things:

1. Will my kid get into a good college?
2. Will my kid get a good job and make money after college?

If you are choosing a career, STEM careers are the gold standard. STEM stands for science, technology, engineering, and math. I've worked at STEM-related day jobs for twenty years and have little trouble finding another job when needed. Job market statistics show salary and opportunity are abundant in STEM fields. Burning Glass, a job-market analytics company, conducted a salary effects study. The following key points will make your jaw drop:

- An average entry-level STEM job had a $15,000 higher salary than a non-STEM job.
- STEM graduates have access to double the entry-level jobs compared to non-STEM graduates.
- The expected growth rate in STEM jobs to non-STEM jobs will increase by 55 percent over the next decade.

Unfortunately, this information has yet to impact decision makers in education. Career counselors are asleep at the wheel. I see minimal encouragement to those interested in a STEM field. Even students focused on other fields have to take STEM courses. Many lessons learned in STEM classes are applicable to all fields. Examples are problem solving, decision making, and system building.

The types of problems in math, business, and systems share the same basic layout:

1. Identify the type of problem at hand.
2. Select the steps needed to come to a solution.

3. Reach your (potential) solution.
4. Evaluate if your solution/answer is viable. Does it make sense? Does it work?
5. If it works, great; otherwise, review another potential solution.

No matter what your field of choice, problem-solving skills are universally applicable. This is why we should encourage our kids to learn and enhance these skills early. With technology moving at the speed of light, opportunity grows by the minute. Your ability to grasp opportunity is based on your problem-solving skills. Sadly, some schools focus on tests and political posturing instead of preparing kids for the REAL WORLD.

I have given talks at a few schools, and here is what I want to see. I want to see high schools inviting back freshmen in college to speak to the senior class in high school. I want to see the same thing with colleges bringing back students who graduated one year ago to talk to the seniors in college.

Speakers discuss what they learned over the last year. Returning college graduates cover topics such as the following:

- Skills and ideas they wish they'd learned senior year in college to be used in the workforce
- Skills and ideas preparing them for the workforce

Returning high school graduates discuss the following:

- Skills and ideas they wish they'd learned senior year in high school to get ahead in college
- Skills and ideas preparing them for college

Imagine the information goldmine this delivers to a student about to take the next big step in their life. If schools adopted this process, students gain insight through foresight. Armed with foresight, students gain a larger advantage when they move up to the next level. The Latin saying applies here: "*Praemonitus praemunitus*," which means, "Forewarned is forearmed."

I think most people on the planet wished they had today's knowledge when they were younger. "If I knew then what I know now, I would be miles ahead of the competition." High school and college are the place to start. Instead, we have kids walking in blind with zero guidance and only test-taking skills and aptitude scores. Rising to the top requires learning speed, which brings me to my next point.

Speed is another motivation for building MathCelebrity for students. In today's marketplace, companies avoid waiting around for a perfect product before releasing it. They build a minimum viable product (MVP) and release it to the market. If it sells, they improve it and release the new version. If it doesn't sell, they scrap it and start over.

Constant innovation, improvisation, and SPEED win this game. When I say "game," I mean the workforce. I mean the business world. I mean financial success. Remember, the market moves at the speed of light. We are handcuffing our kids if we don't teach them to adapt and be nimble with their minds. After college, they will enter the workforce rather naïve.

If you have the time to learn slowly, great. But the real world is moving fast while adaptive learning takes precedence. More and more, you will hear the phrase, "Fail small, fail fast." The name of the game is adapt and improve. What better time to start using this mentality than with our kids in math class?

Problem solving is a career skill used in any career our kids choose. Our kids have fallen behind in this skill set, and we need to shift course to return to greatness.

Everybody involved with raising kids has a duty to start preparing them for the real world—not a test-taking world, not a scholastic etiquette world. I'm talking about a real world where bosses don't give you cupcakes when you miss deadlines. In the real world, your skills need constant updating. In the real world, problems need to be solved. The time is now. Don't leave our kids empty handed.

CHAPTER 26

EPILOGUE

The current time is November 2016. The website continues to grow. We recently smashed our record for most visitors in a day, week, and month. I don't know what the future holds, but I'm excited to push this website as far as humanly possible. My wife has been wonderful support, and I cannot thank her enough. This remains a large project, and her patience has been commendable. She is a great wife and amazing mother to our daughter.

The sales and marketing process is ongoing but worth it. If you sell well, you can write your own ticket. I turned forty this year, and the last thing I want to do is look back and say I missed the window of opportunity. I want my wife and daughter to be proud of me. I want to spend more time with them.

People ask me about daylighting and the corporate world. If you enjoyed those chapters, I have good news for you: I'm writing another book about life in corporate America. I give rules on how a daylighter can navigate through it without being eaten alive. The title is up in the air, but this book is a survival manual for the corporate world. It's a how-to for working on a side job while at a full-time job. I'm excited to share it with you when it's finished.

What comes next is a list of formulas, shortcuts, and tips to help you in math. It covers more than one thousand concepts. It's my version of a math book. For many examples, I lay out the exact method used on MathCelebrity to solve the problem. It is a step up from wordy math book. Instead, it is fluff free and straight to the point. Enjoy.

BASIC MATH

Numerical Properties

Property	Description
Reflexive Property	$A \Leftrightarrow A$
Symmetric Property	If $A = B \Rightarrow B = A$
Transitive Property	If $A = B$ and $B = C \Rightarrow A = C$
Zero Multiplication Property	$A \times 0 = 0$
Zero Property Additive Identity	$A + 0 = 0$
Multiplicative Identity	$A \times 1 = A$
Additive Inverse	$A + (-A) = 0$
Multiplicative Inverse	$A \times \dfrac{1}{A} = 1$
Trichotomy Property	$a < b, a > b$, or $a = b$
Addition Equality	If $a = b \Rightarrow a + c = b + c$
Subtraction Equality	If $a = b \Rightarrow a - c = b - c$
Multiplication Equality	If $a = b \Rightarrow a \times c = b \times c$
Division Equality	If $a = b$ and $c \neq 0 \Rightarrow \dfrac{a}{c} = \dfrac{b}{c}$
Addition Inequality	$\forall\, a, b, c \in \mathbb{R}$ If $a < b \Rightarrow a + c < b + c$
Subtraction Inequality	$\forall\, a, b, c \in \mathbb{R}$ If $a < b \Rightarrow a - c < b - c$

Multiplication Inequality	$\forall\ a, b, c \in \mathbb{R}$ If $a < b \Rightarrow a \times c < b \times c$
Division Inequality	$\forall\ a, b, c \in \mathbb{R}$ If $a < b$ and $c > 0 \Rightarrow a \div c < b \div c$
Associative Property of Addition	$\forall\ a, b, c \in \mathbb{R}$ $(a + b) + c = a + (b + c)$
Associative Property of Multiplication	$\forall\ a, b, c \in \mathbb{R}$ $(a \times b) \times c = a \times (b \times c)$
Commutative Property of Addition	$\forall\ a, b, c \in \mathbb{R}$ $a + b + c = c + b + a$
Commutative Property of Multiplication	$\forall\ a, b, c \in \mathbb{R}$ $a \times b \times c = c \times b \times a$
Distributive Property	$\forall\ a, b, c \in \mathbb{R}$ $a\,(b + c) = a(b) + a(c)$
Fraction Cancellation	$\forall\ a, b, c \in \mathbb{R}$ $b \neq 0$, and $c \neq 0$, $\dfrac{a \times c}{b \times c} = \dfrac{a}{b}$

Basic Math Operations

Addition of 2 numbers
Subtraction of 2 numbers
Multiplication of 2 numbers
Division of 2 numbers
Opposite Numbers
Addition of 3 or more numbers
Multiplication Array

4 rows × 3 columns = 12

$$
\begin{vmatrix}
* & * & * \\
* & * & * \\
* & * & * \\
* & * & *
\end{vmatrix}
$$

Multiplication Tables

	1	2	3	4	5	6	7	8	9	10
1	1	2	3	4	5	6	7	8	9	10
2	2	4	6	8	10	12	14	16	18	20
3	3	6	9	12	15	18	21	24	27	30
4	4	8	12	16	20	24	28	32	36	40
5	5	10	15	20	25	30	35	40	45	50
6	6	12	18	24	30	36	42	48	54	60
7	7	14	21	28	35	42	49	56	63	70
8	8	16	24	32	40	48	56	64	72	80
9	9	18	27	36	45	54	63	72	81	90
10	10	20	30	40	50	60	70	80	90	100

Estimating Sums – Common Core

1. Round each number
2. Add up the rounded numbers

Example: 440 + 35

1. Round 440 $\Rightarrow$ 400
2. Round 35 $\Rightarrow$ 30
3. 400 + 30 = 430

Estimating Reasonableness of Product – Common Core

1. Round each number down
2. Multiply the rounded numbers (a)

3. Take the residual from the first number (Original Number – Rounded Number)
4. Multiply the residual of the first number by the rounded second number (b)
5. Take the residual from the second number (Original Number – Rounded Number)
6. Multiply the residual of the second number by the rounded first number (c)
7. Add up $a + b + c$
8. If the sum is less than the initial estimate, this seems like a valid estimate

Example: 77 × 43 with an estimated product of 3,311. Using the steps above, we have:

1. 70 and 40
2. 70 × 40 = 2,800 (a)
3. 77 – 70 = 7
4. 7 × 40 = 280 (b)
5. 43 – 40 = 3
6. 3 × 70 = 210 (c)
7. 2,800 + 280 + 210 = 3,290
8. Since 3,290 < 3,311, this is a reasonable estimate of the product

Number Bonds

The relationship of a number to the combination of its' parts. We use the number 10 as a marker.

1. Add 8 + 6 using number bonds of 10
2. Since 6 = 2 + 4, we can write 8 + 6 as 8 + 2 + 4
3. Group by 10
4. (8 + 2) + 4

5. 10 + 4

6. 14

Lattice Math Multiplication

14 × 56

1. Write first number across the top and second number down the sides

1	4	
		5
		6

2. Multiply 1 times 5 to get 5 and split the digits in the upper and lower corners
3. Multiply 4 times 5 to get 20 and split the digits in the upper and lower corners
4. Multiply 1 times 6 to get 6 and split the digits in the upper and lower corners
5. Multiply 4 times 6 to get 24 and split the digits in the upper and lower corners

1	4	
0/5	2/0	5
0/6	2/4	6

Add the diagonal totals up: Start at the bottom right and move to the top left:

1. 4 = 4
2. 0 + 6 + 2 = 8
3. 5 + 2 + 0 = 7
4. 0 = 0

Putting together our answer by reading down from the left upper entry to the bottom right, we get 0-7-8-4 = 784

Absolute Value

The absolute value of a number is the distance from a point to the origin on the number line. Anything inside the vertical braces becomes positive.

$$|a| = a$$
$$|-a| = a$$
$$-|-a| = a$$

$$|x| = \begin{cases} x, & \text{if } x \geq 0 \\ -x, & x < 0 \end{cases}$$

Opposite Numbers

The opposite of $a = -a$

The opposite of $-a = -(-a) = a$

Midpoint of a Line

$$\text{Midpoint } \overline{AB} = \frac{a+b}{2}$$

Number Types

Whole – All positive integers including 0 $\{0, 1, 2, 3, \dots\}$
Natural – All positive integers after 0 $\{1, 2, 3, 4, \dots\}$
Even – Numbers divisible by 2 $\{2, 4, 6, 8, \dots\}$
Odd – Numbers not divisible by 2 $\{3, 5, 7, 9, \dots\}$

Prime – Numbers with no other factors than 1 and itself $\{3, 5, 7, 11, \dots\}$
Composite – Numbers with factors other than one and itself $\{4, 6, 8, 10, \dots\}$
Fibonacci Sequence

$$F_n = \begin{cases} 0, & \text{if } n = 0 \\ 1, & \text{if } n = 1 \\ F_{n-1} + F_{n-2}, & \text{if } n > 1 \end{cases} \qquad (1)$$

The first ten Fibonacci numbers are $0, 1, 1, 2, 3, 5, 8, 13, 21, 34$

Algorithm 26.10.1 Fibonacci Algorithm
1: **procedure** FIBONACCI(n)
2:    $f_0 := 0$
3:    $f_1 := 1$
4:    *counter* := 2
5:    **while** *counter* $\leq n$ **do**
6:        $f_n := f_{n-1} + f_{n-2}$
7:           *counter* = *counter* + 1
8:    **end while**
9:    **return** f_n ▷ Return the nth Fibonacci number
10: **end procedure**

Comparing Numbers

Comparing Numbers: Given 2 numbers a and b then 1 of the following 3 is true

If $a < b \Rightarrow a - b < 0$
If $a > b \Rightarrow a - b > 0$
If $a = b \Rightarrow a - b = 0$

Number Operations

Round Truncate Ceiling Floor: Take a number 95.46783

Operation	Explanation	Example
Round	Round a number to a certain amount of digits	Round(95.46783,2) = 95.47
Truncate	Remove all decimals	Truncate(95.46783) = 95
Ceiling	Next highest integer	[95.46783] = 96
Floor	Next lowest integer	]95.46783[= 95
Accuracy	Number of digits to the right of the decimal point	95.46783 has an accuracy of 5
Precision	Number of digits in the number	95.46783 has an precision of 6

Floor of x is also written as $[x]$
Ceiling of x is also written as $[x]$

Tally Marks

Every time you reach a multiple of 5, (5, 10, 15, 20), you use a horizontal line through the tally marks

Number	Tally Marks
1	|
2	||
3	|||
4	||||
5	‖‖
6	‖‖ |
7	‖‖ ||
8	‖‖ |||
9	‖‖ ||||
10	‖‖ ‖‖

11	⊪⊪⊪ ⊪⊪⊪ ⌶
12	⊪⊪⊪ ⊪⊪⊪ ‖
13	⊪⊪⊪ ⊪⊪⊪ ‖|
14	⊪⊪⊪ ⊪⊪⊪ ‖‖
15	⊪⊪⊪ ⊪⊪⊪ ⊪⊪⊪
16	⊪⊪⊪ ⊪⊪⊪ ⊪⊪⊪ ⌶
17	⊪⊪⊪ ⊪⊪⊪ ⊪⊪⊪ ‖
18	⊪⊪⊪ ⊪⊪⊪ ⊪⊪⊪ ‖|
19	⊪⊪⊪ ⊪⊪⊪ ⊪⊪⊪ ‖‖
20	⊪⊪⊪ ⊪⊪⊪ ⊪⊪⊪ ⊪⊪⊪

Count Backwards

Count Backwards

n

$n - x$

$n - 2x$

$n - (n - 1)x$

True False Equations

Determine if the left side of the equation equals the right side of the equation.

Determine if $4 + 1 = 5 + 2$ is true or false.

1. Evaluate the left side: $4 + 1 = 5$
2. Evaluate the right side: $5 + 2 = 7$
3. Since $5 \neq 7$, this is False

Balancing Equations

Balance Addition Equations

Take 4 numbers, n_1, n_2, n_3, n_4

Using addition, we try to balance the equation such that one of the following is true:

$$n_1 + n_2 = n_3 + n_4$$
$$n_1 + n_3 = n_2 + n_4$$
$$n_1 + n_4 = n_2 + n_3$$

Take the numbers 8, 6, 3, 5

$$8 + 6 = 3 + 5 \Rightarrow 14 \neq 8$$
$$8 + 3 = 6 + 5 \Rightarrow 11 = 11$$
$$8 + 5 = 6 + 3 \Rightarrow 13 \neq 9$$

Balance Subtraction Equations

Take 4 numbers, n_1, n_2, n_3, n_4

Using subtraction, we try to balance the equation such that one of the following is true:

$$n_1 - n_2 = n_3 - n_4$$
$$n_1 - n_3 = n_2 - n_4$$
$$n_1 - n_4 = n_2 - n_3$$
$$n_1 - n_3 = n_4 - n_2$$

Take the numbers 5, 6, 2, 9

$5 - 6 = 2 - 9 \Rightarrow -1 \neq -7$
$5 - 2 = 6 - 9 \Rightarrow 3 \neq -3$
$5 - 9 = 6 - 2 \Rightarrow -4 \neq 4$
$5 - 9 = 9 - 6 \Rightarrow -4 = 3$
$9 - 5 = 6 - 2 \Rightarrow 4 = 4$

Balance Multiplication Equations

Take 4 numbers, n_1, n_2, n_3, n_4

Using multiplication, we try to balance the equation such that one of the following is true:

$n_1 \times n_2 = n_3 \times n_4$
$n_1 \times n_3 = n_2 \times n_4$
$n_1 \times n_4 = n_2 \times n_3$

Take the numbers 8, 10, 4, 5

$8 \times 10 = 4 \times 5 \Rightarrow 80 \neq 20$
$8 \times 4 = 10 \times 5 \Rightarrow 32 \neq 50$
$8 \times 5 = 10 \times 4 \Rightarrow 40 = 40$

Balance Division Equations

Take 4 numbers, n_1, n_2, n_3, n_4

Using division, we try to balance the equation such that one of the following is true:

$$\frac{n_1}{n_2} = \frac{n_3}{n_4}$$

$$\frac{n_1}{n_3} = \frac{n_2}{n_4}$$

$$\frac{n_1}{n_4} = \frac{n_2}{n_3}$$

$$\frac{n_1}{n_3} = \frac{n_4}{n_2}$$

Take the numbers 20, 5, 40, 10

$$\frac{20}{5} = \frac{40}{10} \Rightarrow 4 = 4$$

$$\frac{20}{40} = \frac{5}{10} \Rightarrow 0.5 = 0.5$$

$$\frac{20}{10} = \frac{5}{40} \Rightarrow 2 \neq 0.125$$

$$\frac{20}{40} = \frac{10}{5} \Rightarrow 0.5 \neq 2$$

Number Pairs

Number pairs are combinations of two numbers which add up to a given number. See the grid below:

Number	Number Pairs
2	1 + 1
3	1 + 2, 2 + 1
4	1 + 3, 2 + 2, 3 + 1
5	1 + 4, 4 + 1, 2 + 3, 3 + 2
6	1 + 5, 5 + 1, 2 + 4, 4 + 2, 3 + 3
7	1 + 6, 6 + 1, 2 + 5, 5 + 2, 3 + 4, 4 + 3

8	1 + 7, 7 + 1, 2 + 6, 6 + 2, 3 + 5, 5 + 3, 4 + 4
9	1 + 8, 8 + 1, 2 + 7, 7 + 2, 3 + 6, 6 + 3, 4 + 5, 5 + 4
10	1 + 9, 9 + 1, 2 + 8, 8 + 2, 3 + 7, 7 + 3, 4 + 6, 6 + 4, 5 + 5

Decompose Number Pairs

Decomposing a number into pairs is finding all combinations of two numbers that add to our original number. Example: Decompose 5 into number pairs.

$5 = 1 + 4$
$5 = 2 + 3$
$5 = 3 + 2$
$5 = 4 + 1$

Unknown Numbers

Unknown Numbers work like equations using a question mark to identify a missing number. What you need to do is find the number that makes the equation work:

$8 + ? = 11 \Rightarrow ? = 3$
$12 - ? = 7 \Rightarrow ? = 5$

Ten Frames

Dot representation on a basis of 10 per card:

Ten-frame for 5

Ten-frame for 6

Duplication and Mediation

We use this method to determine the product of two numbers. Take 471 × 35

1. Divide the left hand column by 2, and if it is odd, go to the next lowest integer.
2. Double Column 2
3. Cross out any right hand column with an even value in the left hand column
4. Stop when the left hand column equals 1
5. Add up all the right hand columns with an odd value in the left hand column

35	471
$\dfrac{35}{2} = 17$	$471 \times 2 = 942$
$\dfrac{17}{2} = 8$	$942 \times 2 = \cancel{1,884}$
$\dfrac{8}{2} = 4$	$1,884 \times 2 = \cancel{3,768}$
$\dfrac{4}{2} = 2$	$3,768 \times 2 = \cancel{7,536}$
$\dfrac{2}{2} = 1$	$7,536 \times 2 = 15,072$

$471 + 942 + 15,072 = 16,485 \Rightarrow 471 \times 35 = 16,485$

CHAPTER 28

GEOMETRY

Geometry Calculators

2-Dimensional Shape Formulas

Shape	Perimeter	Area	Comments
Circle	$2\pi r$	πr^2	r = radius
Square	$4s$	s^2	s = side
Rectangle	$2l + 2w$	$l \times w$	l = length and w = width
Parallelogram	$2l + 2w$	$l \times w$	l = length and w = width
Equilateral Triangle	$3s$	$\dfrac{1}{2}sh$	s = side and h = height
Isosceles Triangle	$2s + b$	$\dfrac{1}{2}sh$	s = side and h = height and b = base
Kite	$2ss + 2ls$	$d_1 d_2$	ss = short side and ls = long side
Trapezoid	$b_1 + b_2 + s_1 + s_2$	$\dfrac{a(b_1 + b_2)}{2}$	b = base
Pentagon	$5s$	$\dfrac{a \times s}{4}$	s = side

Shape	Perimeter	Area	Comments
Hexagon	$6s$	$2(1+\sqrt{2})s^2$	s = side
Heptagon	$7s$	$\frac{7}{4}s^2\cot\left(\frac{180°}{7}\right)$	s = side
Octagon	$8s$	$2(1+\sqrt{2})s^2$	s = side
Nonagon	$9s$	$\frac{9}{4}s^2\cot\left(\frac{180°}{9}\right)$	s = side
Rhombus	$4s$	$\frac{d_1 d_2}{2}$	s = side, d_1 = Diagonal 1, d_2 = Diagonal 2
Annulus	N/A	$R^2 - r^2$	R = Outer Radius and r = inner radius

3-Dimensional Shape Formulas

Shape	Surface Area	Lateral Area	Volume	Comments
Cube	$6s^2$	$4s^2$	s^3	s = side
Rectangular Solid	$lw + lh + wh$	$h(2l + 2w)$	lwh	l = length, w = width, and h = height
Cylinder	$\pi r^2 h + 2\pi rh$	$2\pi rh$	$\pi r^2 h$	r = radius and h = height
Pyramid	$b \times LA$	$\frac{\text{Base Perimeter} \times \text{Slant Height}}{2}$	$\frac{bh}{3}$	b = base and h = height
Sphere	$4\pi r^2$	N/A	$\frac{4\pi r^3}{3}$	r = radius and h = height
Cone	$\pi r^2 + \pi rl$	N/A	$\frac{\pi r^2 h}{3}$	r = radius and h = height
Torus	$4\pi^2 Rr$	$4\pi^2 Rr$	$(\pi r^2)(2\pi R)$	r = minor radius and R = major radius

Cuboid

Cuboid Measurement	Formula
Volume	$a \times b \times c$
Surface Area	$2(ab + bc + ca)$
Face Diagonal Length dab	$\sqrt{a^2 + b^2}$
Face Diagonal Length dac	$\sqrt{a^2 + c^2}$
Face Diagonal Length dbc	$\sqrt{b^2 + c^2}$
Space Diagonal Length dabc	$\sqrt{a^2 + b^2 + c^2}$

Triangle Items

Angle Ratio for a Triangle $\Rightarrow ax + bx + cx = 180$

Cevian Triangle Items $\Rightarrow a^2n + b^2m = t^2c + mnc$

$$\text{Centroid} = \left(\frac{x_1 + x_2 + x_3}{3}\right), \left(\frac{y_1 + y_2 + y_3}{3}\right)$$

Geometric Mean of a Triangle

$$\frac{\overline{AD}}{\overline{CD}} = \frac{\overline{CD}}{\overline{DB}}$$

Triangle Inequality

$$\triangle ABC \Rightarrow \overline{AB} + \overline{AC} > \overline{BC}, \overline{AB} + \overline{BC} > \overline{AC}, \overline{AC} + \overline{BC} > \overline{AB}$$

Special Triangles

30-60-90 Triangle Side Ratios opposite each angle are $1, \sqrt{3}, 2$

45-45-90 Triangle Side Ratios opposite each angle are $\frac{\sqrt{2}}{2}, \frac{\sqrt{2}}{2}, 1$

Clocks

Clock Angle Calculator

$$\theta H = \frac{60H + M}{2}$$

$\theta m = 6M$

$\Delta = |\theta h - \theta m|$

Clock Hands Meet

1. For $n = 0$ to 11 $\frac{12n}{11}$

2. Hour = Integer Portion

3. Minutes = Integer(60 × Decimal Portion)
4. Seconds = 60 × Decimal Portion

Planar

Eulers Planar Formula $v - e + f = 2$ where v = vertices, e = edges, and f = faces

Polygons

$P = ns$

$$A = \frac{s^2 n}{4 \tan \frac{\pi}{n}}$$

Interior Angle Sum = $(n - 2) \times 180°$

Diagonals = $\dfrac{n(n-3)}{2}$

Vertex Diagonals = $n - 3$

Triangles from one Vertex = $n - 2$

Polygon Sides	Name
3	Triangle
4	Quadrilateral
5	Pentagon
6	Hexatagon
7	Heptagon
8	Octagon
9	Decagon
10	Dodecagon
11 or more	n-gon

Quadrilaterals

Quadrilateral Area: $A = \sqrt{(s-a)(s-b)(s-c)(s-d)}$

Bretschneiders Formula Area:

$$A = \sqrt{(s-a)(s-b)(s-c)(s-d) - abcd \cdot \cos^2(0.5(\alpha + \gamma))a}$$

3 Dimensional Points

Given two 3 Dimensional Points (x_1, y_1, z_1) and (x_2, y_2, z_2):

Distance between the points $D = \sqrt{(x_2 - y_1)^2 + (y_2 - y_1)^2 + (z_2 - z_1)^2}$

Parametric Equations: $(x, y, z) = (x_0, y_0, z_0) + t(a, b, c)$

Symmetric Form Equation: $\dfrac{x - x_0}{a}, \dfrac{y - y_0}{b}, \dfrac{z - z_0}{c}$

Golden Ratio

The Golden Ratio ϕ states: Large Segment A, Small Segment B, and total segment $\overline{AB} \Rightarrow \dfrac{A}{B} = \dfrac{\overline{AB}}{A}$

Identities: $\phi + 1 = \phi^2$ and $\phi - 1 = \dfrac{1}{\phi}$

TRIGONOMETRY

TRIGONOMETRY Calculators

1. Angles and Angle Measurements

Given an angle θ:
Complementary Angle $90 - \theta$
Supplementary Angle $180 - \theta$
Coterminal Angles $\theta \pm 360°$

- Acute Angles when $\theta° < 90$
- Right Angles when $\theta° = 90$
- Obtuse Angles when $\theta° > 90$

2. Degrees Minutes Seconds

Degrees Minutes Seconds to Decimal Degrees $\Rightarrow d° + \dfrac{m}{60} + \dfrac{s}{3600}$
Decimal Degrees to Degrees Minutes Seconds
$$D = Int(d°)$$
$$M = 60 \times (Int(d°) - d°)$$
$$S = 60 \times (M - Int(M))$$

3. Trig Functions

(x, y) is a point other than the origin. Distance from that point to the origin is $r = \sqrt{x^2 + y^2}$

$\sin\theta = \dfrac{y}{r} = \dfrac{\text{opposite}}{\text{hypotenuse}}$

$\cos\theta = \dfrac{x}{r} = \dfrac{\text{adjacent}}{\text{hypotenuse}}$

$\tan\theta = \dfrac{y}{x} = \dfrac{\text{opposite}}{\text{adjacent}}$

$\csc\theta = \dfrac{r}{y} = \dfrac{1}{\sin\theta}$

$\sec\theta = \dfrac{r}{x} = \dfrac{1}{\cos\theta}$

$\cot\theta = \dfrac{x}{y} = \dfrac{1}{\tan\theta}$

4. Pythagorean Theorem

Pythagorean Theorem $a^2 + b^2 = c^2$
$\sin\theta^2 + \cos\theta^2 = 1$
$\sec\theta^2 + \tan\theta^2 = 1$
$\cot\theta^2 + 1 = \csc\theta^2$

5. Cofunctions

For any acute Angle A, the cofunction is listed below:

$\sin A = \cos(90° - A)$
$\cos A = \sin(90° - A)$
$\csc A = \sec(90° - A)$

$$\sec A = \csc(90° - A)$$
$$\tan A = \cot(90° - A)$$
$$\cot A = \tan(90° - A)$$

6. Trig Conversions

$$\text{Radians} = \frac{\theta°\pi}{180}$$

$$\text{Gradians} = \frac{10\theta}{9}$$

Angle Ratio for a triangle with angles a, b and c in the form $a{:}b{:}c = ax + bx + cx = 180$

7. Negative Angle Identities

$$\sin(-\theta) = -\sin(\theta)$$
$$\cos(-\theta) = \cos(\theta)$$
$$\tan(-\theta) = -\tan(\theta)$$
$$\csc(-\theta) = -\csc(\theta)$$
$$\sec(-\theta) = \sec(\theta)$$
$$\cot(-\theta) = -\cot(\theta)$$

8. Product to Sum Identities

For 2 angles, there exists Product to Sum and Sum to product identities

$$\sin(u)\cos(v) = \frac{\sin(u + v) + \sin(u - v)}{2}$$

$$\sin(u)\sin(v) = \frac{\cos(u - v) - \cos(u + v)}{2}$$

$$\cos(u)\cos(v) = \frac{\cos(u-v) + \cos(u+v)}{2}$$

$$\cos(u)\sin(v) = \frac{\sin(a+b) - \sin(a-v)}{2}$$

$$\cos(u+v) = \cos(u)\cos(v) - \sin(u)\sin(v)$$

$$\cos(u-v) = \cos(u)\cos(v) + \sin(u)\sin(v)$$

$$\sin(u+v) = \sin(u)\cos(v) + \cos(u)\sin(v)$$

$$\sin(u-v) = \sin(u)\cos(v) - \cos(u)\sin(v)$$

$$\tan(a+b) = \frac{\tan(a) + \tan(b)}{1 - \tan(a)\tan(b)}$$

$$\tan(a-b) = \frac{\tan(a) - \tan(b)}{1 + \tan(a)\tan(b)}$$

$$\cos(u)\cos(v) = \frac{\cos(u+v) + \cos(u-v)}{2}$$

$$\sin(u)\sin(v) = \frac{\cos(u-v) - \sin(u+v)}{2}$$

$$\sin(u)\cos(v) = \frac{\sin(u+v) + \sin(u-v)}{2}$$

$$\cos(u)\sin(v) = \frac{\sin(u+v) - \cos(u-v)}{2}$$

$$\sin(u) + \sin(v) = 2\sin(\frac{u+v}{2})\cos(\frac{u-v}{2})$$

$$\sin(u) - \sin(v) = 2\cos(\frac{u+v}{2})\sin(\frac{u-v}{2})$$

$$\cos(u) + \cos(v) = 2\cos(\frac{u+v}{2})\cos(\frac{u-v}{2})$$

$$\cos(u) - \cos(v) = 2\sin(\frac{u+v}{2})\sin(\frac{u-v}{2})$$

9. Double Angle Identities

$$\cos(2A) = 1 - 2\sin^2(A)$$
$$\cos(2A) = \cos^2(A) - \sin^2(A)$$
$$\cos(2A) = 2\cos^2(A) - 1$$
$$\sin(2A) = 2\sin(A)\cos(A)$$

$$\tan(2A) = \frac{2\tan(A)}{1 - \tan^2(A)}$$

10. Half Angle Identities

$$\cos\left(\frac{u}{2}\right) = \pm\sqrt{\frac{1 - \cos(u)}{2}}$$

$$\sin\left(\frac{u}{2}\right) = \pm\sqrt{\frac{1 - \cos(u)}{2}}$$

$$\tan\left(\frac{u}{2}\right) = \pm\sqrt{\frac{1 - \cos(u)}{1 + \cos(u)}}$$

$$\tan\left(\frac{u}{2}\right) = \frac{\sin(u)}{1 + \cos(u)}$$

$$\tan\left(\frac{u}{2}\right) = \frac{1 - \cos(u)}{\sin(u)}$$

11. Side and Angle Relations

Side Angle Side

$$A = \frac{1}{2}bc\sin(A)$$

$$A = \frac{1}{2}ab\sin(C)$$

$$A = \frac{1}{2}ac\sin(B)$$

Angle Side Angle

Side Side Side

$$A_1 = \frac{s_2^2 + s_3^2 - s_1^2}{2s_2 s_3}$$

$$A_2 = \frac{s_1^2 + s_3^2 - s_2^2}{2s_1 s_3}$$

$$A_3 = \frac{s_1^2 + s_2^2 - s_3^2}{2s_1 s_2}$$

Law of Sines
$$\frac{a}{\sin(a)} = \frac{b}{\sin(b)} = \frac{c}{\sin(c)}$$

Law of Cosines
$$a^2 = b^2 + c^2 - 2bc \times \cos(A)$$
$$b^2 = a^2 + c^2 - 2ac \times \cos(B)$$
$$c^2 = a^2 + b^2 - 2ab \times \cos(C)$$

12. Herons Formula

Semi-Perimeter $s = \dfrac{a+b+c}{2}$

Area $\Rightarrow A = \sqrt{s(s-a)(s-b)(s-c)}$

13. Arc Length and Sectors of a Circle

Arc Length $\Rightarrow s = r\theta$ where r = radius of the circle and θ is the angle formed by the two lines that connect to the endpoints of the arc

Sector Area $\Rightarrow A = \dfrac{r^2 \theta}{2}$

14. Bearing

Bearing is in the form [NSEW]θ°[NSEW] where the first letter is where you are traveling from and the second letter is where you are going to.

Directional Symbol	Direction
N	North
S	South
E	East
W	West

PRE-ALGEBRA

PRE-ALGEBRA Calculators

1. Place Value and Notation

1. Place Value

The Place Value is as follows for the number 123,456,789

1	2	3	4	5	6	7	8	9
hundred-million	ten-million	million	hundred-thousand	ten-thousand	thousand	hundred	ten	one

For Decimals, the The Place Value is as follows for the decimal 0.12345678

1	2	3	4	5	6	7	8
tenths	hundredths	thousandths	ten-thousandths	hundred-thousandths	millionths	ten-millionths	hundred-millionths

2. Expanded Notation

The number 123,456,789 using expanded notation is as follows:

$$1 \times 10^8 + 2 \times 10^7 + 3 \times 10^6 + 4 \times 10^5 + 5 \times 10^4 + 6 \times 10^3 + 7 \times 10^2 + 8 \times 10^1 + 9 \times 10^0$$

Each number at decimal place (p) n_p is written as $10 \times n_p^{p-1}$

3. Word Notation

Word notation is the verbal expression of a number using the following translations:

hundred-million	ten-million	million	hundred-thousand	ten-thousand	thousand	hundred	ten	one
1	2	3	4	5	6	7	8	9

4. Standard Notation

The number $a(1000) + b(100) + c(10) + d(1)$ is written in standard notation

2. Proportions

Proportions: $\dfrac{a}{b} = \dfrac{c}{d} \Rightarrow a \times d = b \times c$

3. Order of Operations PEDMAS

PEDMAS

P	Parentheses	()
E	Exponent	x^n
D	Division	÷
M	Multiplication	×
A	Addition	+
S	Subtraction	−

4. Factoring and Divisibility

1. Divisibility

Number	Divisibility Rules
2	Last Digit of the number is 0
3	Sum of it's digits ends is divisible by 3
4	Last two digits are divisible by 4
5	The number ends with a 0 or 5
6	If it is divisible by both 2 and 3
7	Multiply each respective digit by 1,3,2,6,4,5 working backwards and repeat as necessary
8	Last 3 digits are divisible by 8
9	Sum of the digits are divisible by 9
10	Last Digit of the number is 0
11	Sum of odd digits Sum of Even Digits is either 0 or Divisible by 11

2. Greatest Common Factor and Least Common Multiple

The Greatest Common Factor is found by taking all of the factors of a group of numbers, finding the common factors, and choosing the highest factor in common of all numbers.

The Least Common Multiple is found by taking all of the multiples of a group of numbers, finding the common multiples, and choosing the highest multiple in common of all numbers.

3. Prime Factorization and Prime Power Decomposition

The prime factorization of a number is expressing a number in terms of the product of primes

The Prime Power Decomposition is the prime factorization grouped by factor.

4. Greatest Common Factor Rewrite Sum

1. Using the distributive property, rewrite $12 + 15$ by factoring out the GCF
2. Find the minimum of your two numbers $\Rightarrow \text{Min}(12,15) = 12$
3. Factors of 12 $\Rightarrow 1, 2, 3, 4, 6, 12$
4. Factors of 15 (up to our minimum of 12) $\Rightarrow 1, 3, 5$
5. Greatest Common Factor of both lists is 3

6. Divide 12 by 3 $\frac{12}{3} = 4$ to get s_1

7. Divide 15 by 3 $\frac{15}{3} = 5$ to get s_2

8. Rewrite using the formula: $GCF(s_1 + s_2)$
9. $3(4+5)$
10. Check our work: $3(4+5) = 3(9) = 27 = 12 + 15$

5. Fractions

1. Operations with common denominators

Operation	Description
Addition	$\dfrac{a}{b} + \dfrac{c}{b} = \dfrac{a+c}{b}$
Subtraction	$\dfrac{a}{b} - \dfrac{c}{b} = \dfrac{a-c}{b}$
Multiplication	$\dfrac{a}{b} \times \dfrac{c}{d} = \dfrac{ac}{bd}$
Division	$\dfrac{a}{b} \div \dfrac{c}{d} = \dfrac{a}{b} \times \dfrac{d}{c} = \dfrac{ad}{bc}$

2. Mixed Number to Improper Fraction

For a mixed number $a\dfrac{b}{c}$, we convert to an improper fraction as follows:

$$\dfrac{c \times a + b}{c}$$

$$1\dfrac{2}{3} \Rightarrow \dfrac{3 \times 1 + 2}{3} \Rightarrow \dfrac{5}{3}$$

3. Reciprocal

The Reciprocal of a fraction $\dfrac{a}{b} = \dfrac{b}{a}$

4. Unit Fraction

For a fraction $\dfrac{a}{b}$, the unit fraction is denoted as $\dfrac{1}{b}$

5. Equivalent Fraction

The Equivalent Fraction of $\frac{a}{b}$ $\Longrightarrow$ Multiply top and bottom by any integer n $\frac{an}{bn}$

6. Percentage and Decimals

Expressing $\frac{a}{b}$ as a Decimal is equivalent to multiplying by 100 percent $\Rightarrow \frac{100 \times a}{b}$

Convert a decimal d with (n) decimal places to a fraction is done by multiplying by $\frac{n \times 10^d}{10^d}$

Convert the percent p to a fraction and decimal $\frac{p}{100}$

7. Interval Counting

$a(b)c$ is read as count a to c in intervals of b

20(10)100 reads count from 20 to 100 in intervals of 10

20, 30, 40, 50, 60, 70, 80, 90, 100

8. Ratios

Ratios

$a : b = \frac{a}{b}$

$$a : b \text{ for } c = \frac{ac}{a+b}$$

$$a : b \text{ and } c : d \Rightarrow \frac{ac}{bd}$$

9. Digit Sum

1. Take a number abc.
2. The digit sum is $a + b + c$
3. If the digit sum is greater than 9, add the digits of the new sum by repeating step 2
4. The reduced digit sum is when the digit sum is less than 10

Take 987654. Digit Sum is $9 + 8 + 7 + 6 + 5 + 4 = 39$
$39 \Rightarrow 3 + 9 = 12$
$12 \Rightarrow 1 + 2 = 3$

10. Exponents

Exponent Rules

$$x^0 = 1$$
$$x^2 = x \cdot x$$
$$x^3 = x \cdot x \cdot x$$
$$x^n = x \cdot x \cdot x \cdots n \text{ times}$$

$$x^{-n} = \frac{1}{x^n}$$

$$b^m \cdot b^n = b^{m+n}$$

$$b^{\frac{m}{n}} = b^{m-n}$$

$$(b^m)^n = b^{mn}$$
$$(bc)^n = b^n c^n$$
$$\left(\frac{b}{c}\right)^n = \frac{b^n}{c^n}$$

1. Simplest Exponent Form

Take $3 \times a \times a \times a \times b \times b$. We want to group all variables that occur more than once using exponents.

We have three a's and two b's. So we take our constant of 3 and then group our variables by exponents.

$$3a^3 b^2$$

11. Square Roots and Roots

$$\sqrt{x^2} = x$$
$$\sqrt[n]{x} = x^{\frac{1}{n}}$$

- $\sqrt{1} = 1$

- $\sqrt{4} = 2$

- $\sqrt{9} = 3$

- $\sqrt{16} = 4$

- $\sqrt{25} = 5$

- $\sqrt{36} = 6$

- $\sqrt{49} = 7$

- $\sqrt{64} = 8$

- $\sqrt{81} = 9$

- $\sqrt{100} = 10$

1. Bakshali Method for Square Roots

1. Find the square root of a number S
2. Starting at 1, square each positive integer until it's total exceeds S. Call the final positive integer N
3. Calculate d $\Rightarrow d = S - N^2$
4. $P = \dfrac{d}{2N}$

5. $A = N + P$

6. $\sqrt{S} \approx A - \dfrac{P^2}{2A}$

2. Newtons Method for Square Roots

1. The square root of a number can be represented by the function
   $$f(x) = x^2 - S$$
2. Taking the Derivative of this, we have $f'(x) = 2x$
3. Since the square root is always positive, we start with $x_0 = 1$
4. $x_1 = x_0 + ((x_0) - S) / f'(x_0)$
5. Repeat by using $x_n = x_{n-1} + ((x_{n-1}) - S) / f'(x_{n-1})$
6. You can stop when the difference between the current and last iteration are close

3. Babylonian Method for Square Roots

Start with $i = 0$, and iterate until the desired result is reached:

$$x_i = \frac{1}{2}\left(x_{i-1} + \frac{S}{x_{i-1}}\right)$$

4. Exponential Identity Method for Square Roots

To find the square root of a number $S \Rightarrow \sqrt{S}$

Evaluate $e^{\frac{Ln(S)}{2}}$

ALGEBRA

ALGEBRA Calculators

1. 1 unknown equations

1. One-Step Equations

One step equations such as $cx = d \Rightarrow 2x = 90$

All you do is take the right side of the equation and divide it by the coefficient of the variable. So $x = \dfrac{d}{c}$

2. Two-Step Equations

Two-step equations such as $ax - b = c \Rightarrow 2x - 9 = 31$

1. Add or subtract to remove the constant from the left hand side
2. Then divide the right hand side of the equation by the coefficient of the variable

For $ax - b = c \Rightarrow x = \dfrac{c + b}{a}$

For $ax + b = c \Rightarrow x = \dfrac{c - b}{a}$

2. Binomials

FOIL – First Outside Inside Last
$$(a + b)(c + d) = (a \times c) + (b \times c) + (a \times d) + (b \times d)$$

Difference of two squares
$$a^2 x^2 - b^2 y^2 = (ax + by)(ax - by)$$

3. Imaginary Numbers

The imaginary Number $i \Rightarrow i = \sqrt{-1}$

$$i^2 = (\sqrt{-1})^2 = -1$$

$$i^3 = (i\sqrt{-1})^2 = -i$$

$$i^4 = ((\sqrt{-1})^2)^2 = (-1)^2 = 1$$

The goal here is to break down the powers. First step is getting power multiples of 4 since $i^4 = 1$

4. Complex Number Operations

Given 2 complex numbers $a + bi$ and $c + di$
Adding: $a + bi + (c + di) = (a + c) + (b + d)i$
Subtracting: $a + bi - (c + di) = (a - c) + (b - d)i$
Multiplying: $(a + bi)(c + di) = ac + adi + bci + bdi^2$

Dividing: $\dfrac{a+bi}{c+di} = \dfrac{(a+bi)(c-di)}{(c+di)(c-di)}$

Square Root: $\sqrt{a+bi}$ has two roots:

$$r = \sqrt{a^2 + b^2}$$

$$y = \sqrt{\frac{1}{2}(r-a)}$$

$$x = \frac{b}{2y}$$

Root 1: $x + yi$
Root 2: $-x - yi$
Absolute Value: $|a+bi| = \sqrt{a^2 + b^2}$

5. Intersection of 2 lines

2 lines are either parallel, perpendicular, or not intersecting.

Parallel $\Rightarrow m_1 = m_2$ where m_1 is the slope of line 1 and m_2 is the slope of line 2

Perpendicular $\Rightarrow m_1 = \dfrac{-1}{m_2}$

If the two lines are not parallel or perpendicular, the lines intersect

The angle between the 2 lines denoted as $\theta \Rightarrow \tan(\theta) = \dfrac{m_2 - m_1}{1 + m_2 m_1}$

6. Quadratic Equations

The solution to the Quadratic Equation $ax^2 + bx + c = 0$ is:

$$x = \frac{-b \pm \sqrt{b^2 - 4ac}}{2a}$$

If the discriminant $b^2 - 4ac > 0$, 2 real unequal roots exist
If the discriminant $b^2 - 4ac = 0$, 1 real root exists
If the discriminant $b^2 - 4ac < 0$, complex conjugate roots exist

Vertex of the parabola formed by the quadratic is $(\frac{-b}{2a}, f(\frac{-b}{2a}))$ where $y = a(x - h)^2 + k$

Concavity is the x^2 coefficient, Up if positive, down if negative.

7. 3 Point Quadratic Equation

Take 3 points: $(x_1, y_1),(x_2, y_2),(x_3, y_3)$

- $a = x_1^2$
- $b = x_1$
- $c = 1$
- $d = y_1$
- $e = x_2^2$
- $f = x_2$
- $g = 1$
- $h = y_2$
- $i = x_3^2$
- $j = x_3$
- $k = 1$
- $l = y_3$

$$\Delta = (a \times f \times k) + (b \times g \times i) + (c \times e \times j) - (c \times f \times i) - (a \times g \times j) - (b \times e \times k)$$

$$a = \frac{(d \times f \times k) + (b \times g \times l) + (c \times h \times j) - (c \times f \times l) - (d \times g \times j) - (b \times h \times k)}{\Delta}$$

$$b = \frac{a \times h \times k) + (d \times g \times i) + (c \times e \times l) - (c \times h \times i) - (a \times g \times l) - (d \times e \times k)}{\Delta}$$

$$c = \frac{(a \times f \times l) + (b \times h \times i) + (d \times e \times j) - (d \times f \times i) - (a \times h \times j) - (b \times e \times l)}{\Delta}$$

8. Descartes Rule of Signs

$$f(x) = 2x^3 - 7x^2 + 4x - 14$$

1. Sign Change + to −
2. Sign Change − to +
3. Sign Change + to −

3 roots − 1 pair (2 roots) = 1

Therefore, we have a possible combination of (3 or 1) positive roots

9. Slope and Line Equations

Slope and Line Equations with 2 points
Given two standard points on a Cartesian Graph of $(x_1, y_1), (x_2, y_2)$

Standard Equation of a line is $y = mx + b$ where $m = \dfrac{y_2 - y_1}{x_2 - x_1}$

Point-Slope Format is $y - y_1 = m(x - x_1)$

Distance between the points is $d = \sqrt{(x_2 - x_1)^2 + (y_2 - y_1)^2}$

Rational Exponents − Fractional Indices

$$x^{\frac{a}{b}} = (\sqrt[b]{x})^a$$

10. System of Equations 2 unknowns

Substitution or Elimination

11. Cramers Rule

1. 2 unknowns

Cramers Rule for 2 unknowns Determinant Δ
Equation 1: $ax + by = c$
Equation 2: $dx + ey = f$
Build the matrix of coefficients

$$\begin{vmatrix} a & c \\ b & d \end{vmatrix}$$

$$\Delta = a \times e + b \times d$$

$$x = \frac{c \times e - b \times f}{\Delta}$$

$$y = \frac{a \times f - c \times d}{\Delta}$$

Augmented matrix $AX = B$

$$A = \begin{vmatrix} a & b \\ d & e \end{vmatrix} \quad X = \begin{vmatrix} x \\ y \end{vmatrix} \quad B = \begin{vmatrix} c \\ f \end{vmatrix}$$

$$A\begin{vmatrix} B \end{vmatrix} = \begin{vmatrix} a & b & | & c \\ d & e & | & f \end{vmatrix}$$

2. 3 unknowns

Cramers Rule for the following system of equations:

$$ax + by + cz = d$$
$$ex + fy + gz = h$$
$$ix + jy + kz = l$$

$$\Delta = (a \times f \times k) + (b \times g \times i) + (c \times e \times j) - (c \times f \times i) - (a \times g \times j) - (b \times e \times k)$$
$$x1 = (d \times f \times k) + (b \times g \times l) + (c \times h \times j) - (c \times f \times l) - (d \times g \times j) - (b \times h \times k)$$
$$y1 = (a \times h \times k) + (d \times g \times i) + (c \times e \times l) - (c \times h \times i) - (a \times g \times l) - (d \times e \times k)$$
$$z1 = (a \times f \times l) + (b \times h \times i) + (d \times e \times j) - (d \times f \times i) - (a \times h \times j) - (b \times e \times l)$$

$$x = \frac{x1}{\Delta}$$

$$y = \frac{y1}{\Delta}$$

$$z = \frac{z1}{\Delta}$$

Augmented matrix $AX = B$

$$A = \begin{vmatrix} a & b & c \\ e & f & g \\ i & j & k \end{vmatrix} \quad X = \begin{vmatrix} x \\ y \\ z \end{vmatrix} \quad B = \begin{vmatrix} d \\ h \\ l \end{vmatrix}$$

$$A \mid B = \begin{vmatrix} a & b & c & \mid & d \\ d & e & f & \mid & g \\ h & i & j & \mid & k \end{vmatrix}$$

3. 4 unknowns

Cramers Rule for the following system of equations:

$$aw + bx + cy + dz = e$$
$$fw + gx + hy + iz = j$$
$$kw + lx + my + nz = 0$$

$$pw + qx + ry + sz = t$$

$$\Delta_1 = (agms) + (ahnq) + (ailr) - (aimq) - (agnr)$$

$$\Delta_2 = -(ahls) - (fbms) - (fcnq) - (fdlr) + (fdmq)$$

$$\Delta_3 = (fbnr) + (fcls) + (kbhs) + (kciq) + (kdgr)$$

$$\Delta_4 = -(kdhq) - (kbir) - (kcgs) - (pbhn) - (pcil)$$

$$\Delta_5 = -(pdgm) + (pdhl) + (pbim) + (pcgn)$$

$$\Delta = \Delta_1 + \Delta_2 + \Delta_3 + \Delta_4 + \Delta_5$$

$$W1 = (egms) + (ehnq) + (eilr) - (eimq) - (egnr)$$

$$W2 = -(ehls) - (jbms) - (jcnq) - (jdlr) + (jdmq)$$

$$W3 = (jbnr) + (jcls) + (obhs) + (ociq) + (odgr)$$

$$W4 = -(odhq) - (obir) - (ocgs) - (tbhn) - (tcil)$$

$$W5 = -(tdgm) + (tdhl) + (tbim) + (tcgn)$$

$$W = W1 + W2 + W3 + W4 + W5$$

$$X1 = (ajms) + (ahnt) + (aior) - (aimt) - (ajnr)$$

$$X2 = -(ahos) - (fems) - (fcnt) - (fdor) + (fdmt)$$

$$X3 = (fenr) + (fcos) + (kehs) + (kcit) + (kdjr)$$

$$X4 = -(kdht) - (keir) - (kcjs) - (pehn) - (pcio)$$

$$X5 = -(pdjm) + (pdho) + (peim) + (pcjn)$$

$$X = X1 + X2 + X3 + X4 + X5$$

$$Y1 = (agos) + (ajnq) + (ailt) - (aioq) - (agnt)$$

$$Y2 = -(ajls) - (fbos) - (fenq) - (fdlt) + (fdoq)$$

$$Y3 = (fbnt) + (fels) + (kbjs) + (keiq) + (kdgt)$$

$$Y4 = -(kdjq) - (kbit) - (kegs) - (pbjn) - (peil)$$

$$Y5 = -(pdgo) + (pdjl) + (pbio) + (pegn)$$

$$Y = Y1 + Y2 + Y3 + Y4 + Y5$$

$$Z1 = (agmt) + (ahoq) + (ajlr) - (ajmq) - (agor)$$

$$Z2 = -(ahlt) - (fbmt) - (fcoq) - (felr) + (femq)$$

$$Z3 = (fbor) + (fclt) + (kbht) + (kcjq) + (kegr)$$

$$Z4 = -(kehq) - (kbjr) - (kcgt) - (pbho) - (pcjl)$$

$$Z5 = -(pegm) + (pehl) + (pbjm) + (pcgo)$$

$$Z = Z1 + Z2 + Z3 + Z4 + Z5$$

$$w = \frac{W}{\Delta}$$

$$x = \frac{X}{\Delta}$$

$$y = \frac{Y}{\Delta}$$

$$z = \frac{Z}{\Delta}$$

Augmented matrix $AX = B$

$$A = \begin{vmatrix} a & b & c & d \\ f & g & h & i \\ k & l & m & n \\ p & q & r & s \end{vmatrix} \quad X = \begin{vmatrix} w \\ x \\ y \\ z \end{vmatrix} \quad B = \begin{vmatrix} e \\ j \\ o \\ t \end{vmatrix}$$

$$A \,\big|\, B = \begin{vmatrix} a & b & c & d & | & e \\ f & g & h & i & | & j \\ k & l & m & n & | & o \\ p & q & r & s & | & t \end{vmatrix}$$

12. Cubic Equations

$$ax^3 - bx^2 - cx + d = 0$$

$$\Delta = 4b^3 d - b^2 c^2 + 4ac^3 - 18abcd + 27a^2 d^2$$

$$f = \frac{(3c\,/\,a) - b^2\,/\,a^2}{3}$$

$$g = \frac{2b^3\,/\,a^3 - 9bc\,/\,a^2 + 27d\,/\,a}{27}$$

$$h = \frac{g^2}{4} + \frac{f^3}{27}$$

$$i = \sqrt{0.25g^2 - h}$$

$$j = i^{1/3}$$

$$k = \text{Arccosine}\left(\frac{-g}{2i}\right)$$

$$l = -j$$

$$m = \text{Cos}\left(\frac{k}{3}\right)$$

$$n = \sqrt{3} \times \text{Sin}\left(\frac{k}{3}\right)$$

$$p = \frac{-b}{3a}$$

Real Roots are (x_1, x_2, x_3)

$$x_1 = 2j \times \text{Cos}\left(\frac{k}{3}\right) - \frac{b}{3a}$$

$$x_2 = l(m+n) + p$$

$$x_3 = l(m-n) + p$$

13. Quartic Equations

$$ax^4 + bx^3 - cx^2 - dx + e = 0$$

$$f = c - \frac{3b^2}{8}$$

$$g = d + \frac{3b^3}{8} - \frac{bc}{2}$$

$$h = e - \frac{3b^4}{256} + \frac{b^2c}{16} - \frac{bd}{4}$$

Using our f, g, and h values, we form the following cubic equation:

$$x^3 + \frac{fx^2}{2} + \frac{f^2 - 4h}{16}x - \frac{g^2}{64} = 0$$

Then, use the cubic formula above to break down the cubic into roots

14. Scientific Notation

The Scientific Notation for a number is denoted as $a \times (10^n)$ where $1 \le a < 10$

15. Intercepts

The X and Y intercepts for $ax + by = c$

Y Intercept, isolate Y, and set x to 0

$$by = c - ax \Rightarrow \frac{by}{b} = \frac{c - a(0)}{b} \Rightarrow y = \frac{c}{b}$$

X Intercept, isolate X, and set y to 0

$$ax = c - by \Rightarrow \frac{ax}{a} = \frac{c - b(0)}{a} \Rightarrow x = \frac{c}{a}$$

16. Expanding Polynomials

1. Binomial Theorem

Expanding
$$(a+b)^n = a^n + nP_1 a^{n-1}b + nP_2 a^{n-2}b^2 + \cdots + nP_k a^{n-k}b^k + \cdots + b^n$$

17. Factoring Polynomials

Difference of Cubes Factoring $a^3 - b^3 = (a-b)(a^2 + ab + b^2)$
Sum of Cubes Factoring $a^3 + b^3 = (a+b)(a^2 - ab + b^2)$

18. Synthetic Division

Synthetic Division is used to determined roots.

19. Variation Equations

1. Direct Variation Equations

Variation Equation Type	Relation Equation
Direct Variation	$y = kx$
Inverse Variation	$y = \dfrac{k}{x}$
Squared Variation	$y = kx^2$
Cubed Variation	$y = kx^3$
Square Root Variation	$y = k\sqrt{x}$
Inverse of the Square Variation	$y = \dfrac{k}{x^2}$
Inverse of the Cube Variation	$y = \dfrac{k}{x^3}$
Inverse of the Square Root Variation	$y = \dfrac{k}{\sqrt{x}}$

2. Joint Variation Equations

Variation Equations Example:

a varies jointly with b and c, and $a = 12$ when $b = 1$ and $c = 6$, solve for a when $b = 2, c = 3$ with a relationship of $a = kbc$

PRE-CALCULUS

PRE-CALCULUS

The Antilog of a using base b is b^a

Base Conversions

Conjugates

To simplify an expression such as $\dfrac{1}{a+\sqrt{b}}$ we multiply the numerator

and denominator by the conjugate $a - \sqrt{b}$

$$\frac{a-\sqrt{b}}{(a+\sqrt{b})(a-\sqrt{b})} = \frac{a-\sqrt{b}}{a^2 - b}$$

1. Demoivre's Theorem

Demoivres Theorem

<u>Basic</u>

If $z = r\mathrm{cis}(\theta) \Rightarrow z^n = r^n\mathrm{cis}(n\theta)$

<u>Polar</u>

$z = r(\cos(\theta) + i\sin(\theta))$ where

$a = r\cos(\theta)$

$b = r\sin(\theta)$

$$r = \sqrt{a^2 + b^2}$$

2. Logarithms

Logarithms

1. Properties

If $b^y = x \Rightarrow \log_b(x) = y$

$\log_b xy = \log_b x + \log_b y$

$\log_b\left(\dfrac{x}{y}\right) = \log_b x - \log_b y$

$\log_b(x) = \dfrac{\log_x}{\log_b}$

Change of Base

$\log_b(x) = \dfrac{\text{Ln}(x)}{\text{Ln}(b)}$

2. Natural Log

$\text{Ln}(e^x) = e^{\text{Ln}(x)} = x$

$\text{Ln}(x^n) = n \times \text{Ln}(x)$

$\text{Ln}(xy) = \text{Ln}(x) + \text{Ln}(y)$

$\text{Ln}\left(\dfrac{x}{y}\right) = \text{Ln}(x) - \text{Ln}(y)$

3. Conics

1. Circles

Standard Form Equation of a circle is $(x - h)^2 + (y - k)^2 = r^2$
General Form Equation of a circle is
$x^2 + y^2 - 2hx - 2ky + h^2 + k^2 - r^2 = 0$
where (h, k) is the center of the circle and r is the radius.
Chord Length = $2\sqrt{r^2 - t^2}$ where r = radius and t = Circle Center to Chord Midpoint Distance

2. Hyperbolas

Standard Equation of a Horizontal Hyperbola is $\dfrac{(x-h)^2}{a^2} + \dfrac{(y-k)^2}{b^2} = 1$

Asymptotes are $y = \pm \dfrac{b}{a} x$

Standard Equation of a Vertical Hyperbola is $\dfrac{(y-k)^2}{a^2} + \dfrac{(x-h)^2}{b^2} = 1$

Asymptotes are $y = \pm \dfrac{a}{b} x$

Foci at $(0, c)$ and $(0, -c)$ where $c = \sqrt{a^2 + b^2}$

Eccentricity: $\varepsilon = \dfrac{\sqrt{a^2 - b^2}}{a}$

Latus Rectum: $LR = \dfrac{2b^2}{a}$

3. Parabolas

General equation of the Parabola is $Ax^2 + Cy^2 + Dx + Ey + F = 0$

Vertex at (h, k)

Standard Vertical Parabola Equation is $(x - h)^2 = 4c(y - k)$

$$\begin{cases} c > 0, & \text{opens up if } c > 0 \\ c < 0, & \text{opens down if } c < 0 \end{cases} \qquad (2)$$

Focus at $(0, c)$ and directrix at $y = c$

Standard Horizontal Parabola Equation is $(y - c)^2 = 4c(x - h)$ then we have

$$\begin{cases} c > 0, & \text{opens right if } c > 0 \\ c < 0, & \text{opens left if } c < 0 \end{cases} \qquad (3)$$

Focus at $(c, 0)$ and directrix at $x = c$

4. Ellipse

Standard Equation of a Horizontal Ellipse is $\dfrac{(x - h)^2}{a^2} + \dfrac{(y - k)^2}{b^2} = 1$

Standard Equation of a Vertical Ellipse is $\dfrac{(x - h)^2}{b^2} + \dfrac{(y - k)^2}{a^2} = 1$

5. Polar Conics

The Vertical Directrix Polar Conic Equations where e = eccentricity, d = directrix are denoted as:

$$r = \frac{ed}{1 - e(cos(\theta))}$$

$$r = \frac{ed}{1 + e(cos(\theta))}$$

The Horizontal Directrix Polar Conic Equations where e = eccentricity, d = directrix are denoted as:

$$r = \frac{ed}{1 - e(sin(\theta))}$$

$$r = \frac{ed}{1 + e(sin(\theta))}$$

4. Counting Formulas

Factorials $n! = n \times (n-1) \times (n-2) \times \cdots \times 1$

Algorithm 31.4.1 Factorial Algorithm
1: **procedure** FACTORIAL(n)
2:    *answer*:= *n*
3:    **while** *n* > *0* **do**   ▷ Keep multiplying until we hit 1
6:       *n*:= *n – 1*
7:       *answer*:= *answer* × *n*
8:    **end while**
9:    **return** *answer*   ▷ The factorial is the positive integer answer
10: **end procedure**

Stirling Approximation

$$n! \approx \sqrt{2\pi n}\left(\frac{n}{e}\right)^n$$

1. Permutations

The Permutations formula for choosing k ways from n possibilities is

$$_nP_k = \frac{n!}{(n-k)!}$$

With permutations, order matters.

2. Combinations

The Combinations formula for choosing k unique ways from n possibilities is

$$_nC_k = \frac{n!}{k!(n-k)!}$$

With combinations, order does not matter.

Symmetry property: $_nC_k =_n C_{nk}$

Pascals identity: $_{n+1}C_k =_n C_{k1} +_n C_k$

3. Group Combinations

Group Combinations are found from multiplying each individual combination. A Group/Team consist of 5 men and 7 women. How many Group/Teams of 5 can be formed using 3 men and 2 women?

$$C(5,3) \times C(7,2)$$

5. Functions

Function Test
For all the (x, y) pairs, a function exists if there are not more than one x value
Function Test
$$f(x) = ax^n + bx^{n-1} + \cdots + c$$

CALCULUS

CALCULUS

1. Quadrants

Quadrant	(x, y)
I	(x, y)
II	(x, y)
III	$(-x, -y)$
IV	$(x, -y)$

2. Point Reflection and Rotation

Reflection Type	Reflected Point
y-axis	$(x, y) \Rightarrow (-x, y)$
x-axis	$(x, y) \Rightarrow (x, -y)$
origin	$(x, y) \Rightarrow (-x, -y)$

Point Rotation

Rotation Degrees	New Point
$90°$	$(x, y) \Rightarrow (-y, x)$
$180°$	$(x, y) \Rightarrow (-x, -y)$
$270°$	$(x, y) \Rightarrow (y, -x)$

3. Sequences and Series

1. Summation Properties

$$\sum_{i=1}^{n} (a_i + b_i) = \sum_{i=1}^{n} a_i + \sum_{i=1}^{n} b_i$$

$$\sum_{i=1}^{n} ca_i = c \sum_{i=1}^{n} a_i$$

$$\sum_{i=1}^{n} 1 = n$$

2. Series Formulas

Series Type	Explicit Formula
Arithmetic	$a_n = a_1 + (n-1)d$
Geometric	$a_n = a_1(r - n)$
Infinite Geometric	$\sum_{k=1}^{\infty} ar^{k-1} = \dfrac{a}{1-r}$
Harmonic	$a_n = \dfrac{1}{n}$

4. Limits

Constants: $\lim_{x \to a} c = c$

Single Variable: $\lim_{x \to a} x = a$

Coefficients: $\lim_{x \to a} cf(x) = c \lim_{x \to a} f(x)$

Single Variable with Exponent: $\lim_{x \to a} x^n = a^n$

Addition (Limit of a sum is the sum of the limits): $\lim_{x \to a} [f(x) + g(x)] = \lim_{x \to a} f(x) + \lim_{x \to a} g(x)$

Subtraction (Limit of a difference is the difference of the limits): $\lim_{x \to a} [f(x) - g(x)] = \lim_{x \to a} f(x) - \lim_{x \to a} g(x)$

Multiplication (Limit of a product is the product of the limits): $\lim_{x \to a} [f(x) \cdot g(x)] = \lim_{x \to a} f(x) \cdot \lim_{x \to a} g(x)$

Division (Limit of a quotient is the quotient of the limits):

$$\lim_{x \to a} \left[\frac{f(x)}{g(x)} \right] = \frac{\lim_{x \to a} f(x)}{\lim_{x \to a} g(x)}$$

Roots: $\lim_{x \to a} \sqrt[n]{x} = \sqrt[n]{a}$

Two Sided Limit:
$\lim_{x \to a} f(x) = L \Leftrightarrow \lim_{x \to a^-} f(x) = L = \lim_{x \to a^+} f(x)$

Squeeze Theorem or Sandwich Theorem: If $f(x) \le g(x)$ when x is near a and the limits of $f(x)$ and $g(x)$ both exist as x approaches a, then:

$\lim_{x \to a} f(x) = \lim_{x \to a} h(x) = L$ then $\lim_{x \to a} g(x) = L$

1. Continuity

A function f is continuous at a number a if the following three conditions hold:

1. $f(a)$ is defined
2. $\lim_{x \to a} f(x)$ exists
3. $\lim_{x \to a} f(x) = f(a)$

A function is continuous from the right at a number a if $\lim_{x \to a^+} f(a)$

A function is continuous from the left at a number a if $\lim_{x \to a^-} f(a)$

The following types of functions are continuous at every number in their domains:

- Polynomials
- Rational functions
- Root functions
- Trigonometric functions
- Inverse trigonometric functions
- Exponential functions
- Logarithmic functions

$$\lim_{x \to a} f(g(x)) = f(\lim_{x \to a} g(x))$$

if g is continuous at a and f is continous at $g(a)$, the the composite function $f \circ g$ given by $(f \circ g)(x) = f(g(x))$ is continous at a

2. Intermediate Value Theorem

If f is continuous on the closed interval $[a,b]$, then let N be any number between $f(a)$ and $f(b)$. A number c exists in (a,b) such that $f(c) = N$

3. Vertical Asymptote Limit

The line $x = a$ is a vertical asymptote of the curve $y = f(x)$ if one or more of the following statements is true:

- $\lim_{x \to a} f(x) = \infty$
- $\lim_{x \to a^-} f(x) = \infty$
- $\lim_{x \to a^+} f(x) = \infty$
- $\lim_{x \to a} f(x) = -\infty$
- $\lim_{x \to a^-} f(x) = -\infty$
- $\lim_{x \to a^+} f(x) = -\infty$

4. Horizontal Asymptote Limit

The line $y = L$ is a horizontal asymptote of the curve $y = f(x)$ if either:

- $\lim_{x \to \infty} f(x) = L$
- $\lim_{x \to -\infty} f(x) = L$

5. Other Limit Laws

$\lim_{x \to \infty} \frac{1}{x^n} = 0$

$\lim_{x \to -\infty} \frac{1}{x^n} = 0$

$\lim_{x \to \infty} e^x = 0$

6. Eulers Constant Limit

- $e \approx 2.7182818$
- $\lim_{x \to 0} (1+x)^{\frac{1}{x}} = e$
- $e = \lim_{n \to \infty} (1 + \frac{1}{n})^n$

5. Rates of Change

6. Tangent to the curve

$y = f(x)$ at the point $P(a, f(a))$ is a line through P with slope:

$$m = \lim_{x \to a} \frac{f(x) - f(a)}{x - a}$$

With $h = x - a$ and $x = a + h$, we have:

$$m = \frac{f(a+h) - f(a)}{h}$$

7. Instantaneous Rate of Change

$y = f(x)$ and x changes from x_1 to x_2

Then the change, or increment in x is: $\Delta x = x_2 - x_1$ and corresponding change in y is $\Delta y = f(x_2) - f(x_1)$

Average rate of change of y with respect to x: $\dfrac{\Delta y}{\Delta x} = \dfrac{f(x_2) - f(x_1)}{x_2 - x_1}$

Instantaneous rate of change: $\lim_{\Delta x \to 0} \dfrac{\Delta y}{\Delta x} = \lim_{x_2 \to x_1} \dfrac{f(x_2) - f(x_1)}{x_2 - x_1}$

8. Derivatives

Definition

$$f'(x) = \lim_{\delta x \to 0} \frac{(f(x + \Delta x) - f(x))}{\Delta x}$$

Constant Rule: $f(x) = c \Rightarrow f'(x) = 0$

Power Rule with no coefficient: $f(x) = x^n \Rightarrow f'(x) = nx^{n-1}$

Power Rule with coefficient: $f(x) = ax^n \Rightarrow f'(x) = nax^{n-1}$

Exponential General: $f(x) = e^x \Rightarrow f'(x) = e^x$

Exponent Special: $f(u) = e^u \Rightarrow f'(u) = e^u \, du$

Chain Rule 1: $(f(x)$ and $g(x)$ both differentiable: $F(x) = f \cdot g(x) \Rightarrow F'(x) = f'(g(x))g'(x)$

Chain Rule 2: $y = f(u)$ and $u = g(x) \to \dfrac{dy}{dx} = \dfrac{dy}{du}\dfrac{du}{dx}$

Product Rule: $y = f(x)g(x) \Rightarrow y' = f(x)g'(x) + f'(x)g(x)$

Quotient Rule: $y = \dfrac{f(x)}{g(x)} \Rightarrow y' = \dfrac{(f(x)g(x) - g(x)f(x))}{g^2}$

Power Rule and Chain Rule (n is a real number and u is differentiable:

$$f(x) = u^n \Rightarrow f'(x) = nu^{n-1}u'$$

1. Derivative Rules

- A function is differentiable at a if $f(a)$ exists
- If f is differentiable at a, then f is continuous at a
- If $f'(x) > 0$ on an interval, then f is increasing on that interval
- If $f'(x) < 0$ on an interval, then f is decreasing on that interval
- If $f''(x) > 0$ on an interval, then f is concave upward on that interval
- If $f''(x) < 0$ on an interval, then f is concave downward on that interval

2. Trigonometric function derivatives

- $f(x) = \sin(x) \Rightarrow f'(x) = \cos(x)$
- $f(x) = \cos(x) \Rightarrow f'(x) = -\sin(x)$
- $f(x) = \tan(x) \Rightarrow f'(x) = \sec^2(x)$
- $f(x) = \csc(x) \Rightarrow f'(x) = -\csc(x) \cdot \cot(x)$
- $f(x) = \sec(x) \Rightarrow f'(x) = \sec(x) \cdot \tan(x)$
- $f(x) = \cot(x) \Rightarrow f'(x) = -\csc^2(x)$

3. Parametric Curve Tangents

Parametric equation curves: $x = f(t), y = g(t)$

$$\frac{dy}{dt} = \frac{dy}{dx} \cdot \frac{dx}{dt}$$

$$\frac{dy}{dx} = \frac{\dfrac{dy}{dt}}{\dfrac{dx}{dt}}$$

4. Inverse Trigonometric function derivatives

- $f(x) = \sin^{-1}(x) \Rightarrow f'(x) = \dfrac{1}{\sqrt{1-x^2}}$

- $f(x) = \cos^{-1}(x) \Rightarrow f'(x) = \dfrac{-1}{\sqrt{1-x^2}}$

- $f(x) = \tan^{-1}(x) \Rightarrow f'(x) = \dfrac{1}{1+x^2}$

- $f(x) = \csc(x) \Rightarrow f'(x) = \dfrac{-1}{|x|\sqrt{x^2-1}}$

- $f(x) = \sec(x) \Rightarrow f'(x) = \dfrac{1}{|x|\sqrt{x^2-1}}$

- $f(x) = \cot(x) \Rightarrow f'(x) = \dfrac{-1}{1+x^2}$

5. Logarithmic function derivatives

- $f(x) = \log_a x \Rightarrow f'(x) = \dfrac{1}{x \cdot \text{Ln}(a)}$

- $f(x) = \text{Ln}(x) \Rightarrow f'(x) = \dfrac{1}{x}$

- $f(x) = \text{Ln}(u) \Rightarrow f'(x) = \dfrac{1}{u}\dfrac{du}{dx}$

- $f(x) = \text{Ln}(|x|) \Rightarrow f'(x) = \dfrac{1}{x}$

9. Hyperbolic Functions

Hyperbolic sine: $\sinh(x) = \dfrac{e^x - e^{-x}}{2}$

Hyperbolic cosine: $\cosh(x) = \dfrac{e^x + e^{-x}}{2}$

Hyperbolic tangent: $\dfrac{\sinh(x)}{\cosh(x)} = \dfrac{1 - e^{-2x}}{1 + e^{-2x}}$

Hyperbolic secant: $\dfrac{1}{\cosh(x)} = \dfrac{2}{e^x + e^{-x}}$

Hyperbolic cosecant: $\dfrac{1}{\sinh(x)} = \dfrac{2}{e^x - e^{-x}}$

Hyperbolic cotangent: $\dfrac{1}{\tanh(x)} = \dfrac{e^x + e^{-x}}{e^x - e^{-x}}$

10. Taylor Polynomial

n-th degree polynomial of f centered at a:

$$T_n(x) = f(a) + f'(a)(x - a) + \frac{f''(a)}{2}(x - a) + \ldots + \frac{f'^n(a)}{2}(x - a)^n$$

11. Maximum and Minimum

- A function f has a local maximum or local minimum at c if $f(c) \geq f(x)$ when x is near c
- Fermat's Theorem: If f has a local maximum at c and if $f'(c)$ exists, then $f'(c) = 0$
- If f has a local maximum or minimum at c, then c is a critical number of f

12. L'Hospitals Rule

If f and g are differentiable and $g'(x) \neq 0$ near a:

- $\lim_{x \to a} f(x) = 0$

- $\lim_{x \to a} g(x) = 0$

- $\lim_{x \to a} f(x) = \pm\infty$

- $\lim_{x \to a} g(x) = \pm\infty$

then $\lim_{x \to a} \dfrac{f(x)}{g(x)} = \lim_{x \to a} \dfrac{f'(x)}{g'(x)}$

Summary: Limit of a quotient of functions equals limit of the quotient of derivatives

13. Integrals

Definition $\int_a^b f(x)dx = \lim_{n \to \infty} \sum_{i=1}^{n} f(x_i)\Delta x$

Power Rule $\int ax^n dx = \dfrac{ax^{n+1}}{n+1}$

STATISTICS

1. Probability

1. Probability Axioms

- We start with a sample space denoted Ω or S, which is the set of all possible outcomes
- $P(\Omega) = 1$
- We have an event, which is a subset of a sample space
- Probability is assigned to an event, which is the chance that occurs in the entire sample space
- $P(\varnothing) = 0$
- We have nonnegativity of probabilities, so for an event A, we have $P(A) \geq 0$
- Since the sample space is the set of all possible outcomes, we have $P(\Omega) = 1$
- Complement: The probability of anything but A is $P(A^c) = 1 - P(A) \Rightarrow P(A) = 1 - P(A^c)$
- Combine the sample space, one event and an event complement, we have $P(\Omega) = P(A \cup A^c) = 1$

- Additivity: if $A \cap B = \emptyset$, then $P(A \cup B) = P(A) + P(B)$
- if $A \subset B, B = A \cup (A^c \cap B)$
- $A \subset B \Rightarrow P(A) \le P(B)$
- $A = A \cap (B \cup B^c) = (A \cap B) \cup (A \cap B^c)$

2. Bayes Rule and Conditional Probability

Bayes Rule $P(B \mid A) = \dfrac{P(A \cap B)}{P(A \cap B) + P(A \cap B^c)}$

$A_1, A_2, ..., A_n$ are sets in S where $A_i \cap A_j = \emptyset$ for $i \ne j$

$$P(B) = P(B \cap A_1) + P(B \cap A_2) + \cdots + P(B \cap A_n)$$

$$P(B \mid A) = \frac{P(A \mid B) \times P(B)}{P(A \mid B) \times P(B) + P(A \mid B^c) \times P(B^c)}$$

$$P(A_i \cap B) = P(B \mid A_i) \cdot P(A_i)$$

If A and B are mutually exclusive, they cannot occur at the same time. Also, they are not independent, so:

Event	Independent Events Probability	Mutually Exclusive Events Probability
$P(A \cap B)$	$P(A) \cdot P(B)$	0
$P(B \mid A)$	$P(B)$	0
$P(A \mid B)$	$P(A)$	0

3. Bayes Theorem Table Method (3 Events)

A_i	$P(A_i)$	$P(B \mid A_i)$	$P(A_i \cap B)$	$P(A_i \mid B)$
A_1				
A_2				
A_3				
Total	1.0			

4. Fundamental Rule of Counting

Fundamental Rule of Counting states that the total number of ways one can do something is with k items is $n_1 \cdot n_2 \cdot \ldots \cdot n^k$

5. De Morgan Laws

• $(A \cap B)^C = A^C \cup B^C$
• $(A \cup B)^C = A^C \cap B^C$

2. Venn Diagram

The Venn Diagram is a way to show the intersection of 2 or more events. The white area below represents $A \cap B$

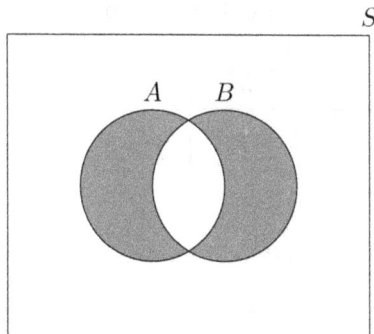

1. Law of Total Probability

$$P(B) = P(B \cap A_1) + P(B \cap A_2) + \cdots + P(B \cap A_n) = \sum_{i=1}^{n} P(B \cap A_i)$$

Two Event Example: $P(B) = P(A \cap B) + P(A^C \cap B)$

$$P(B) = P(A) \cdot P(B \mid A) + P(A^C) \cdot P(B \mid A^C)$$

2. Event Probabilities

Postulate I: $P(A) \geq 0$

Postulate II: Sample Space S, $P(S) = 1$

Postulate III (Mutually Exclusive Events):
$P(A_1 \cup A_2 \cup PA_3 \cup \cdots) = P(A_1) + P(A_2) + P(A_3) + \cdots$

Empty Set: $P(\varnothing) = 0$ for Sample Space S

$$P(E) = \frac{\text{Favorable Outcomes}}{\text{Total Outcomes}}$$

3. Tabular Display

Tabular Display to test if a valid sample space. $\sum_{x=1}^{n} p_x = 1$

x	x_1	x_2	x_3	$\cdots$	x_n
$p(x)$	$p(x_1)$	$p(x_2)$	$p(x_3)$	$\cdots$	$p(x_n)$

Expected value: $E(x) = x_1 p(x_1) + x_2 p(x_2) + x_3 p(x_3) + \cdots + x_n p(x_n)$

4. 2 Events

2 Number Sets: $P(A \cup B) = P(A) + P(B) - P(A \cap B)$
$(A \cap B) \subset (A \cup B)$
If A and B are mutually exclusive (never happen together), then:
$P(A \cap B) = 0$

5. 3 Events

3 Events:
$$P(A \cup B \cup C) = P(A) + P(B) + P(C) - P(A \cap C)$$
$$-P(B \cap C) + P(A \cap B \cap C)$$
$$A \cup (B \cap C) = (A \cup B) \cap (A \cup C)$$
$$A \cap (B \cup C) = (A \cap B) \cup (A \cap C)$$

6. Odds Probability

The Odds Probability in Favor for m successes out of n is $\dfrac{m}{m+n}$

The Odds Probability Against for m successes out of n is $\dfrac{n}{m+n}$

7. Probability Distribution of Discrete Random Variable

Probability Distribution: $f(x) = P(X = x)$. The function must satisfy two conditions:

1. $f(x) \geq 0$ for each value in the domain
2. $\Sigma f(x) = 1$

Probability mass function (PMF) must satisfy two conditions:

1. $0 \le p_X(x) \le 1$
2. $\Sigma p_X(x) = 1$

8. Probability Density of Continuous Random Variable

Probability Density Function: $P(a \le X \le B) = \int_a^b f(x)dx$. The function must satisfy two conditions:

For a continuous random variable X, $P(X = a) = 0$

1. $f(x) \ge 0$ for $-\infty < x < \infty$ for each value in the domain

2. $\int_{-\infty}^{\infty} f(x)dx = 1$

3. $\int_a^a f(x)dx = 0 \Rightarrow$ probability of any one point is zero

We also have: $P(a \le X \le B) = F(b) - F(a)$

Continuous Conditional probability: $f_{X|Y}(x \mid y) = \dfrac{f_{X|Y}(x, y)}{f_Y(y)}$

9. Cumulative Distribution Function (CDF)

1. Non-decreasing function, since as x grows larger, $P(X \le x)$ increases
2. $F_x(-\infty) = 0$
3. $F_x(\infty) = 1$

$F(x) = P(X \le x)$
X is a random variable, x is an algebraic variable
Survival function looks at the complement of the CDF
$S(x) = S_X(x) = P(X > x) = 1 - F(x)$

10. Discrete Random Variable Probability Mass Function

- Discrete means the range, or possible outcomes, is a finite countable set
- Discrete Variable Probability Mass Function (PMF): $p(x) = p_X(x) = P(X = x)$ $0 \le p(x) \le 1$
- $\Sigma p(x) = 1$

11. Mixed Distribution

- CDF: $F_X(x) = \alpha_1 F_1(x) + \alpha_2 F_2(x) + \cdots + \alpha_n F_n(x)$ where $\alpha_1 + \alpha_2 + \cdots + \alpha_n = 1$
- If all X_i are continuous, we have: $f_X(x) = \alpha_1 f_1(x) + \alpha_2 f_2(x) + \cdots + \alpha_n f_n(x)$
- Moment Generating Function: $M_X(t) = \alpha_1 M_1(t) + \alpha_2 M_2(t) + \cdots + \alpha_n M_n(x)$

12. Expectation

1. Expected Value of a Random Variable

Discrete Random Variable $E(X) = \sum\limits_{k=x}^{n} x \cdot f(x)$

Continous Random Variable $\int_{-\infty}^{\infty} x \cdot f(x)$

Continuous Random Variable Shortcut: Since the survival function is $S(x) = 1 - F(x)$, we have $E(x) = \int_0^{\infty} S(x)dx$ if the function is non-negative.

Corollary 1, a is a constant: $E(aX) = aE(X)$

Corollary 2, c is a constant: $E(c) = c$

This gives us: $E(aX + b) = aE(x) + b$

Variance: $\text{Var}(aX + b) = a^2 \cdot \text{Var}(X)$

Variance of a constant c: $\text{Var}(c) = 0$

2. Moments

Discrete Random Variable $E(X^r) = \sum\limits_{x}^{n} x^r \cdot f(x)$

Continuous Random Variable $E(X^r) = \sum\limits_{-\infty}^{\infty} x^r \cdot f(x)dx$

Variance: $\mathrm{Var}(X) = E[(X-\mu)^2] = E(X^2) - (E(X))^2$

Standard Deviation: $\sigma = \sqrt{\mathrm{Var}(x)}$

3. Chebyshevs Theorem

The Probability that x is within k standard deviations of the mean is denoted as:
$$P(|X-\mu| < k\sigma) \geq 1 - \frac{1}{k^2}$$

4. Moment Generating Function

Discrete Random Variable: $M_x(t) = E(e^{tx}) = \sum\limits_{x}^{n} e^{tx} \cdot f(x)$

Continuous Random Variable: $M_x(t) = E(e^{tx}) = \sum\limits_{-\infty}^{\infty} e^{tx} \cdot f(x)dx$

Two requirements for a valid moment generating function:

1. $M_x(0) = 1$
2. Variance is greater than 0, so $E(X^2) - (E(X))^2 > 0$

If a and b are constants, then: $M_{aX+b}(t) = e^{bt} \cdot M_X(at)$
If X and Y are independent, then: $M_{X+Y}(t) = M_X(t) \cdot M_Y(t)$

Coefficient of Variation: $CV(X) = \dfrac{\sigma}{\mu} = \dfrac{SD(X)}{E(X)}$

13. Special Probability Distributions

1. Discrete Statistical Distributions

Uniform Measurement	Formula
Mean $\bar{x}$	$\dfrac{\sum\limits_{i=1}^{n} x_i}{n}$
Median (Odd Number of Values)	$x_{\frac{n+1}{2}}$
Median (Even Number of Values)	$0.5 \times \left(x_{\frac{n}{2}} + x_{\frac{n+1}{2}} \right)$
Mode	Highest Occurring Number
Variance σ^2	$\dfrac{\sum\limits_{i=1}^{n} (x_i - \bar{x})^2}{n}$
Standard Deviation σ	$\sqrt{\sigma^2}$
Skewness	$\dfrac{(\Sigma x_i - \bar{x})^3}{(n-1)\sigma^3}$
Kurtosis	$\dfrac{(\Sigma x_i - \bar{x})^4}{((\Sigma x_i - \bar{x})^2)^2}$
Standard Error of the Mean	$\dfrac{\sigma}{\sqrt{n}}$
Average Deviation	$\dfrac{\mid x_i - \bar{x} \mid}{n}$
Range	$x_{i\max} - x_{i\min}$
Mid-Range	$\dfrac{x_{i\max} - x_{i\min}}{2}$
Entropy	$\text{Ln}(n)$
Harmonic Mean	$\dfrac{N}{\frac{1}{x_1} + \frac{1}{x_2} + \ldots + \frac{1}{x_n}}$
Geometric Mean	$(x_1 \times x_2 \times \ldots \times x_n)^n$

Distribution	Mean	Variance	Probability Formula	Moment Generating Function $M_X(t)$
Bernoulli	np	$p(1-p)$	$p^k q^{n-k}$	$(1-p) + e^{tp}$
Binomial	np	$np(1-p)$	$n! p^k q^{n-k}$	$[1 + np(e^t - 1)]^n$
Negative Binomial	$\dfrac{k}{p}$	$\dfrac{k(1-p)}{p^2}$	$\dfrac{(n-1)! p^k q^{n-k}}{(k-1)!(n-k)!}$	
Poisson	λ	λ	$\dfrac{\lambda^k}{e^\lambda k!}$	$e^{\lambda(e^t - 1)}$
Geometric	$\dfrac{1}{p}$	$\dfrac{1-p}{p^2}$	pq^{n-1}	
Hypergeometric	$\dfrac{nk}{N}$	$\dfrac{nk(N-k)(N-n)}{N^2(N-1)}$	$\dfrac{(_k C_x) \times (_{N-k} C_{n-x})}{_N C_n}$	
Multinomial	np_i	$np_i(1-p_i)$	$\dfrac{n!}{x_1! \cdots x_k!} p_1^{x_1} \cdots p_k^{x_k}$	$(\sum\limits_{i=1}^{k} p_i e^{ti})^n$

The mode is the value that maximizes the mass or density function
$\Rightarrow f'(x) = 0$

2. Quartiles

Quartile	Percent or Formula
First (Lower)	25th percentile
Second (Median)	50th percentile
Third (Upper)	75th percentile
Interquartile Range (IQR)	75th – 25th percentile

$F_X(x) = p$ where p is the percentile in decimal form
Decile goes in 10 percents

3. Special Probability Densities

Distribution	PDF	Mean	Variance
Uniform	$\dfrac{1}{b-a}$	$\dfrac{a+b}{2}$	$\dfrac{1}{12(b-a)^2}$
Exponential	$\lambda e^{-\lambda x}$	$\dfrac{1}{\lambda}$	$\dfrac{1}{\lambda^2}$
Normal Distribution	$\dfrac{1}{\sigma\sqrt{2\pi}}e^{\frac{1}{2}\left(\frac{x-\mu}{\sigma}\right)^2}$	μ	σ^2

Normal Approxiation to the Binomial Distribution is $Z = \dfrac{X - n\theta}{\sqrt{n\theta(1-\theta)}}$

14. Sampling Distributions

1. Sample Formulas

Sample Mean $\overline{X} = \dfrac{\sum_{i=1}^{n} X_i}{n}$

Sample Variance $S^2 = \dfrac{\sum_{i=1}^{n}(X_i - \overline{X})^2}{n-1}$

2. Central Limit Theorem

Given a random sample from an infinite population with mean μ and variance σ^2, the limiting distribution is

$$Z = \dfrac{\overline{X} - \mu}{\dfrac{\sigma}{n}}$$

3. Law of Total Expectation

$E(X) = E(E(X \mid Y))$
Law of Total Variance: $\mathrm{Var}(X) = E(\mathrm{Var}(X \mid Y)) \mid \mathrm{Var}(E(X \mid Y))$

4. Independent Variables

If X and Y are independent, then: $E(XY) = E(X) \cdot E(Y)$

5. Chi-Square

$\dfrac{(n-1)S^2}{\sigma^2}$ has a chi-square distribution with n - 1 degrees of freedom.

Chi Squared where O_i is observed values and E_i is expected value

$$\chi_c^2 = \Sigma \frac{(O_i - E_i)^2}{E_i}$$

6. t-distribution

If Y has a chi-square distribution with v degrees of freedom, Z has a normal distribution, then the Student t Distribution of $T = \dfrac{Z}{\sqrt{Y/v}}$ is

$$f(t) = \frac{\Gamma \frac{(v+1)}{2}}{\sqrt{\pi v}\, \Gamma \frac{v}{2}} \cdot (1 + \frac{t^2}{v})^{-\frac{v+1}{2}} \text{ for } -\infty < t < \infty$$

7. F-distribution

F Distribution
When U and V are independent random variables having chi-square distributions with v_1 and v_2 degrees, of freedom, then:

$$F = \frac{\frac{U}{v_1}}{\frac{V}{v_2}}$$

is a random variable having an F-distribution

15. Decision Theory

1. Game Theory

Payoff Matrix

PlayerB/PlayerA	a_1	a_2
θ_1	$L(a_1, \theta_1)$	$L(a_2, \theta_1)$
θ_2	$L(a_1, \theta_2)$	$L(a_2, \theta_2)$

where L represents the loss function

16. Estimation Theory

$\widehat{\Theta}$ is an unbiased estimator of θ if $E(\widehat{\Theta}) = \theta$

1. Maximum Likelihood

$$L(\theta) = f(x_1, x_2, \cdots, x_n; \theta)$$

17. Confidence Intervals

1. Confidence Interval of the Mean

Confidence Interval for a Large Sample for the Mean

$$\bar{x} - z_{\frac{\alpha}{2}} \cdot \frac{\sigma}{\sqrt{n}} < \mu < \bar{x} + z_{\frac{\alpha}{2}} \cdot \frac{\sigma}{\sqrt{n}}$$

Confidence Interval for a Small Sample for the Mean

$$\bar{x} - t_{\frac{\alpha}{2}, n-1} \cdot \frac{s}{\sqrt{n}} < \mu < \bar{x} + t_{\frac{\alpha}{2}, n-1} \cdot \frac{s}{\sqrt{n}}$$

2. Confidence Interval of the Difference Between Means

Confidence Interval for a Large Sample for the Difference of Means

$$(x_1 - x_2) - z_{\frac{\alpha}{2}} \cdot \sqrt{\frac{\sigma_1^2}{n_1} + \frac{\sigma_2^2}{n_2}} < \mu_1 - \mu_2 < (x_1 - x_2) + z_{\frac{\alpha}{2}} \cdot \sqrt{\frac{\sigma_1^2}{n_1} + \frac{\sigma_2^2}{n_2}}$$

Confidence Interval for a Small Sample for the Difference of Means

$$(x_1 - x_2) - t_{\frac{\alpha}{2}, n_1 + n_2 - 2} \cdot s_p \sqrt{\frac{1}{n_1} + \frac{1}{n_2}} < \mu_1 - \mu_2 < (x_1 - x_2) + t_{\frac{\alpha}{2}, n_1 + n_2 - 2} \cdot s_p \sqrt{\frac{1}{n_1} + \frac{1}{n_2}}$$

3. Confidence Interval for Estimation of Proportions

Confidence Interval for a Proportion:

$$\hat{\theta} - z_{\frac{\alpha}{2}} \cdot \sqrt{\frac{\hat{\theta}(1 - \hat{\theta})}{n}} < \theta < \hat{\theta} + z_{\frac{\alpha}{2}} \cdot \sqrt{\frac{\hat{\theta}(1 - \hat{\theta})}{n}}$$

4. Confidence Interval for Difference of Proportions

Confidence Interval for the Difference of Proportions:

$$(\hat{\theta}_1 - \hat{\theta}_{12}) - z_{\frac{\alpha}{2}} \cdot \sqrt{\frac{\hat{\theta}_{11}(1-\hat{\theta}_{11})}{n_1} + \frac{\hat{\theta}_{12}(1-\hat{\theta}_1)}{n_2}} < (\theta_1 - \theta_2) < \hat{\theta}_1 - \hat{\theta}_2 + z_{\frac{\alpha}{2}}$$

$$\cdot \sqrt{\frac{\hat{\theta}_{11}(1-\hat{\theta}_{11})}{n_1} + \frac{\hat{\theta}_{12}(1-\hat{\theta}_{12})}{n_2}}$$

5. Confidence Interval of the Variance

Confidence Interval for Variance:

$$\frac{(n-1)s^2}{\chi^2_{\frac{\alpha}{2},\, n-1}} < \sigma^2 < \frac{(n-1)s^2}{\chi^2_{1-\frac{\alpha}{2},\, n-1}}$$

18. Hypothesis Testing

1. Hypothesis Errors

- Type I Error – Reject the null hypothesis when it is true. This probability equals α
- Type II Error – Accept the null hypothesis when it is false. This probability equals β

2. Neyman-Pearson Lemma

$$L_0 = \prod_{i=1}^{n} f(x_i; \theta_0) \text{ and } L_1 = \prod_{i=1}^{n} f(x_i; \theta_1)$$

3. Power Function of a Test

For H_0 against H_1

$$\pi(\theta) = \begin{cases} \alpha(\theta) \\ 1 - \beta(\theta) \end{cases}$$

4. Tests Concerning Means

$H_0 : \mu = c$
$H_A : \mu \neq c$

Test Statistic: $Z = \dfrac{\bar{x} - \mu}{\dfrac{\sigma}{\sqrt{n}}}$

Critical Value (CV): Reject H_0 if $z \leq -CV$ or $z \geq CV$

5. Tests for the Difference of Means

$H_0 : \mu_1 - \mu_2 = 0$
$H_A : \mu_1 - \mu_2 > 0$

Test Statistic

$$t = \frac{\bar{x}_1 - \bar{x}_2 - \sigma}{s_p \sqrt{\dfrac{1}{n_1} + \dfrac{1}{n_2}}} \quad \text{where } s_p = \frac{(n_1 - 1)s_1^2 + (n_2 - 1)s_2^2}{n_1 + n_2 - 2}$$

6. Tests for Proportions

$H_0 : p = c$
$H_A : p \neq c$

Test Statistic

$$\hat{p} = \frac{x}{n}$$

$$z = \frac{\hat{p} - p}{\sqrt{\frac{p(1-p)}{n}}}$$

7. Proportion Sample Size

If the proportion $\hat{p}$ is known

$$n = \frac{z^2 \times \hat{p} \times (1 - \hat{p})}{SE^2} \quad \text{where } \hat{p} = \frac{p}{P}$$

If the proportion $\hat{p}$ is NOT known

$$n = \frac{z^2 \times 0.25}{SE^2}$$

8. Goodness of Fit

Test Statistic is $\chi^2 = \frac{(f_i - e_i)^2}{e_i}$

$$\bar{x} = \frac{\Sigma xf}{n}$$

$$s^2 = \frac{\Sigma (x - \bar{x})^2 f}{n-1}$$

19. Regression and Correlation

Covariance and Regression

$$COV(X,Y) = \frac{\Sigma(X_i - \bar{X})(Y_i - \bar{Y})}{n}$$

$$COV(X,Y) = E(XY) - E(X) \cdot E(Y)$$

Correlation Coefficient $r = \dfrac{COV(X,Y)}{s_x s_y}$

$$COV(X,Y) = E[(X - \mu_x)(Y - \mu_y)]$$

Coefficient of Determination $= r^2$

Least Squares Regression Line $\hat{y} = \alpha + \beta x$

$$\beta = \frac{\Sigma(X_i - \bar{X})(Y_i - \bar{Y})}{\Sigma(X_i - \bar{X})^2}$$

1. Covariance Properties

- $\text{Cov}(X,Y) = \text{Cov}(Y,X) \Rightarrow$ Symmetry
- $\text{Cov}(X,X) = \text{Var}(X)$
- $\text{Cov}(aX,Y) = a \cdot \text{Cov}(X,Y)$
- If a is a constant, then $COV(a,X) = 0$
- If a and b are constants, then $COV(a,b) = 0$
- If a and b are constants, then $COV(aX,bY) = ab \cdot COV(X,Y)$
- If a and b are constants, then $COV(X+a,Y+b) = COV(X,Y)$
- If a and b are constants, then
  $\text{Var}(aX + bY) = a^2 \cdot \text{Var}(X) + 2ab \cdot COV(X,Y) + b^2 \cdot \text{Var}(Y)$
- If X and Y are independent, then $COV(X,Y) = 0$
- $\text{Cov}(X,Y)^2 \leq \text{Var}(X) \cdot \text{Var}(Y) \Rightarrow$ Cauchy-Schwartz Inequality

20. Joint Moment Generating Functions

$$M_{X,Y}(s,t) = E(e^{sX+tY})$$

21. Mean Absolute Percentage Error (MAPE)

1. Take a set of actual values $(a_1, a_2, \cdots, a_n)$
2. Take a set of forecasted values $(f_1, f_2, \cdots, f_n)$

3. For each a_i and f_i, we calculate $\dfrac{|a_i - f_i|}{|a_i|}$

4. Sum up all the values in Step 3

5. MAPE = $\dfrac{100\% \times \text{Sum}}{n}$

22. Odds Ratio

Given two events:

- Event X with success probability p_x and failure probability $1 - p_x = q_x$
- Event Y with success probability p_y and failure probability $1 - p_y = q_y$

We have the following grid:

Event	$Y = 1$	$Y = 0$
$X = 1$	p_x	q_x
$X = 0$	p_y	q_y

We have the following conditional probabilities:

Event	$Y = 1$	$Y = 0$
$X = 1$	$\dfrac{p_{11}}{(p_{11} + p_{10})}$	$\dfrac{p_{10}}{(p_{11} + p_{10})}$
$X = 1$	$\dfrac{p_{01}}{(p_{01} + p_{00})}$	$\dfrac{p_{00}}{(p_{01} + p_{00})}$

Our odds ratio is $\dfrac{p_{11} \times p_{00}}{p_{10} \times p_{01}}$

23. Fishers Exact Test

The Fisher exact test table setup below:

Item	Column 1	Column 2	Row Total
Row 1	a	b	$a + b$
Row 2	c	d	$c + d$
Column Total	$a + c$	$b + d$	$n = (a + b + c + d)$

Probability $(p) = \dfrac{(a+b)!(c+d)!(a+c)!(b+d)!}{a!b!c!d!n!}$

24. Fishers Transformation

Given r, we have: $a = \dfrac{1+r}{1-r}$

$z = \dfrac{\text{Ln}(a)}{2}$

25. Fishers Inverse

Given z, we have: $r = \dfrac{e^{2z} - 1}{e^{2z} + 1}$

26. Grand Mean

The Grand Mean (GM) is the mean of means.

1. Take (m) number sets
2. Take the mean of each number set n_i
3. Add up all the means: $N = n_1 + n_2 + \cdots + n_i$
4. $GM = \dfrac{N}{m}$

27. Trimmed Mean

Given a number set with (n) components and a trimmed mean percentage (p), we have:

1. $g = \lfloor p \times n \rfloor$
2. We remove the bottom (g) entries and the top (g) entries from the original number set to get a trimmed number set (t).
3. Our trimmed number set now has $n - 2g$ entries
4. Calculate the mean of the trimmed number set $\dfrac{\Sigma t_i}{n - 2g}$

28. Winsorized Mean

Given a number set with (n) components and a winsorized mean percentage (p), we have:

1. $g = \lfloor p \times n \rfloor$
2. We remove the bottom (g) entries and the top (g) entries from the original number set.

3. We replace them with the next closest entries for a new number set w

4. The new number set still has (n) entries

5. Calculate the mean of the new number set $\dfrac{\Sigma w_i}{n}$

29. Missing Average

Given a set of scores $(s_1, s_2, \cdots, s_n)$ and a target average amount (A), the missing value (m) to meet that average is:

$$m = n \times A - \Sigma s_n$$

30. Coin Toss Probability

$$P(H) = \frac{1}{2} = 0.5$$

$$P(T) = \frac{1}{2} = 0.5$$

31. Event Likelihood

An event (e) is a valid probability is $0 \le P(e) \le 1$

32. Mcnemar Test

The Mcnemar table setup below:

Item	Column 1	Column 2	Row Total
Row 1	a	b	$a + b$
Row 2	c	d	$c + d$
Column Total	$a + c$	$b + d$	$n = (a + b + c + d)$

Set up hypothesis test:

$$H_0 : p_b = p_c$$
$$H_1 : p_b \neq p_c$$

Test Statistic: $\chi^2 = \dfrac{(b-c)^2}{b+c}$

Calculate critical value and if test stat is in rejection region, reject H_0

33. Point Estimate and Margin of Error

Given a lower bound l, an upper bound u, and a sample size n, calculate the point estimate (PE) and margin of error (MOE).

$$PE = \frac{u+l}{2}$$

$$MOE = u - PE$$

34. Analysis of Variance (ANOVA)

1. One-Way Analysis of Variance (ANOVA)

$$SST = SS(Tr) + SSE$$

$$f = \frac{MS(Tr}{MSE}$$

Source of Variation	Degrees of Freedom	Sum of Squares	Mean Square	f
Treatments	$k-1$	$SS(Tr)$	$MS(Tr)$	$\dfrac{MS(Tr)}{MSE}$
Error	$k(n-1)$	SSE	MSE	
Total	$kn-1$	SST		

2. Two-Way Analysis of Variance (ANOVA)

Source of Variation	Degrees of Freedom	Sum of Squares	Mean Square	f
Treatments	$k - 1$	$SS(\text{Tr})$	$MS(\text{Tr})$	$\dfrac{MS(\text{Tr})}{MSE}$
Blocks	$n - 1$	SSB	MSB	$\dfrac{MSB}{MSE}$
Error	$(n - 1)(k - 1)$	SSE	MSE	
Total	$kn - 1$	SST		

35. Nonparametric Test

Sign Test Large Sample

$$\mu = n\theta$$
$$\sigma = \sqrt{n\theta(1 - \theta)}$$

$$Z = \frac{x - \mu}{\sigma}$$

36. Rule of Succession

Rule of Succession s successes in n independent trials, what is the probability that the next repetition is a success

$$P(X_{n+1} \mid X_1 + \ldots + X_n = s) = \frac{s+1}{n+2}$$ where s is the number of successes and n is the number of trials

FINANCE

Calculators

1. Interest and Discount

Interest and Discount Relationships	Variables
$v = \dfrac{1}{1+i}$	v = discount factor and i = interest rate
$d = \dfrac{i}{1+i}$	v = discount factor and i = interest rate
$d = iv$	v = discount factor and i = interest rate
$v = 1 - d$	v = discount factor and d = rate of discount
$\sigma = \mathrm{Ln}(1+i)$	σ = force of interest and i = interest rate
$i = \dfrac{d}{1-d}$	i = interest rate and d = rate of discount

1. Simple and Compound and Continuous Interest

Accumulated Value using Simple Interest $\Rightarrow AV = P(1+it)$
Accumulated Value using Compound Interest $\Rightarrow AV = P(1+i)^t$
Accumulated Value using Continuous Interest $\Rightarrow AV = Pe^{rt}$
where i = interest rate, t = time, e = Euler's constant, and r = rate

2. Simple and Compound Discount

Accumulated Value using Simple Discount $\Rightarrow AV = \dfrac{P}{1-dt}$

Accumulated Value using Compound Discount $\Rightarrow AV = \dfrac{P}{(1-d)^t}$

where d = rate of discount and t = time

2. Interest Applications

1. Rule of 72

The Rule of 72 states the time (n) for money to double using an interest rate of (i) is $n \approx \dfrac{0.72}{i}$

Exact Time for Money to double $\Rightarrow n = \dfrac{\text{Ln}(2)}{\text{Ln}(1+i)}$

When given an interest rate, make sure it is in percentage form, so 6 percent is 0.06 as a decimal, but we use 6. $\dfrac{72}{6} = 12$

3. Annuities

Annuity Calculation	Formula
Present Value of Annuity Immediate	$PV = \text{Payment} \times \dfrac{1-v^n}{i}$
Present Value of Annuity Due	$PV = \text{Payment} \times \dfrac{1-v^n}{d}$
Present Value of Continuous Annuity Immediate	$PV = \text{Payment} \times \dfrac{1-v^n}{\sigma}$
Accumulated Value of Annuity Immediate	$AV = \text{Payment} \times \dfrac{(1+i)^n-1}{i}$

Accumulated Value of Annuity Due	$AV = \text{Payment} \times \dfrac{(1+i)^n - 1}{d}$
Accumulated Value of Continuous Annuity Immediate	$AV = \text{Payment} \times \dfrac{(1+i)^n - 1}{\sigma}$
Present Value of Geometric Annuity Immediate	$\text{Payment} \times \dfrac{1 - \dfrac{1 + k^n}{1 + i}}{i - k}$
Accumulated Value of Geometric Annuity Immediate	$\text{Payment} \times \dfrac{(1+i)^n - (i+k)^n}{i - k}$

4. Perpetuities

Method	Formula	Notes
Present Value of Perpetuity Immediate	$PV = \dfrac{P}{i}$	P = Payment, i = interest rate, payment due at the end of the period
Present Value of Perpetuity Due	$PV = \dfrac{P}{d}$	P = Payment, i = interest rate, payment due at the beginning of the period
Present Value of Arithmetic Perpetuity Immediate	$PV = \dfrac{P_1}{i} + \dfrac{AP}{i^2}$	P = Payment, i = interest rate, payment due at the end of the period
Present Value of Arithmetic Perpetuity Due	$PV = \dfrac{P_1}{d} + \dfrac{AP}{d^2}$	P = Payment, A = Arithmetic Amount, i = interest rate, payment due at the beginning of the period

5. Yield Rates

1. Time Weighted Interest

$$\text{Time Weighted Interest} \quad \frac{MV_1 - MV_0 + D_1 - CF}{MV_0}$$

2. Dollar Weighted Interest

Dollar Weighted Interest

$$I = \frac{2 \times (\text{Invest Income} - \text{Invest Expenses})}{I} (\text{Asset} + \text{Asset} + \text{Income} - \text{Expenses})$$

3. Effective Yield

Compounding Period	n
Daily	365
Weekly	52
Semi-monthly	24
Monthly	12
Quarterly	4
Semi-annually	2
Annually	1

Nominal Interest Rate = i_n

Effective interest rate $i_e = \dfrac{i_n}{n}$

Yield Rate = $(1 + i_e)^n - 1$

4. Nominal Yield

Compounding Period	n
Daily	365
Weekly	52
Semi-monthly	24
Monthly	12
Quarterly	4
Semi-annually	2
Annually	1

$$n_r = n \times e^{(\text{Ln}(\text{EffectiveRate}+1)/n)} - 1$$

5. Inflation

The relationship between Nominal Interest, Real Interest, and Inflation is $1 + n = (1 + i)(1 + r)$

6. Amortization

1. Mortgage

The Mortgage formula is $P = \dfrac{L \times APR}{1 - (\dfrac{1}{1 + APR})^n}$

The Interest Only Mortgage formula is $P = \dfrac{L \times APR}{12}$

Loan Piece	Symbol	Formula
Discount Factor	v	$\dfrac{1}{1 + APR}$
Interest Paid at time t	I_t	$P(1 - v^{n-t+1})$
Principal Paid at time t	P_t	$P - I_t$
Balance at time t	B_t	$\dfrac{P(1 - v^{n-t})}{APR}$

2. Loan with Varying Series of Payments

A loan (L) with a series of (n) periodic installments payment at time (t) of R_t:
$$L = \sum_{i=1}^{n} v^t R_t$$

3. Sinking Fund

The Sinking Fund Payment = Deposit (D) + Interest (I) where:

$$\frac{\text{Loan} \times i}{(1+i)^n - 1} \quad \text{and} \quad I = L \times i$$

7. Depreciation Formulas

Depreciation Method	Depreciation at time t D_t	Book Value at time t B_t
Straight Line	$\dfrac{A - S}{n}$	$\left(1 - \dfrac{t}{n}\right) \times A + S \times \dfrac{t}{n}$
Declining Balance	$dA(1 - d)^{t-1}$	$A(1 - d)^t$
Double Declining Balance	$\dfrac{200 \times B_{t-1}}{n}$	$A - \Sigma B_t$
Sum of the Years Digits	$\dfrac{(A - S)(N - t + 1)}{\Sigma_{x \to 0} n}$	Book Value at time t B_t
Sinking Fund	$\dfrac{(A - S)(1 + j)^{t-1}}{s_n}$	N/A
Units of Output	$\dfrac{(A - S)(\text{PeriodUnits})}{\text{ProductionUnits}}$	N/A

The Capitalized Cost and Periodic Charge: $H = Ai + D + M$ where
Ai is the loss of interest on the original purchase price of the Asset (A)
D is the Depreciation Expense using a Sinking Fund Method
M is the Maintenance Expense

8. Interest Rate Approximation Formulas

Method	Formula
Maximum Yield	$\dfrac{2 \times \text{Yearly Payments} \times \text{Finance Charge}}{\text{Loan} \times (n+1) - \text{Finance Charge} \times (n-1)}$
Minimum Yield	$\dfrac{2 \times \text{Yearly Payments} \times \text{Finance Charge}}{\text{Loan} \times (n+1) + \text{Finance Charge} \times (n-1)}$
Constant Ratio	$\dfrac{2 \times \text{Yearly Payments} \times \text{Finance Charge}}{L(n+1)}$
Direct Ratio	$\dfrac{2 \times \text{Yearly Payments} \times \text{Finance Charge}}{L(n+1) + \dfrac{1}{3} \times \text{Finance Charge} \times (n-1)}$

9. Bonds

1. Bond Price Methods

- P = Price of the Bond
- F = Par Value of Face Amount
- C = Redemption Value
- r = coupon rate
- Fr = coupon amount
- g = modified coupon rate
- i = yield rate
- n = number of periods from calculation date to redemption date
- K = Present Value
- G = Base Amount = Fr/i

Bond Price Method	Bond Price Formula
Basic	$P = \text{Face} \times \text{Coupon} \times \dfrac{1 - v^t}{y} + \dfrac{R}{1 + y^t}$
Premium/Discount	$P = R + (\text{Face} \times \text{Coupon} - R \times \text{yield}) \times \dfrac{1 - v^t}{\text{yield}}$
Base	$P = \dfrac{\text{Face} \times \text{Coupon}}{\text{yield}} + (R - \dfrac{\text{Face} \times \text{Coupon}}{\text{yield}})(1 + \text{yield})^t$
Makeham	$P = \dfrac{R}{(1 + \text{yield})^t} + \dfrac{\text{Face} \times \text{Coupon}}{\text{Yield} \times R} \times (R - \dfrac{R}{(1 + \text{yield})^t})$
Zero Coupon	$P = \dfrac{F}{(1 + \text{yield})^t}$

2. Bond Yield Methods

Bond Yield Method	Yield Rate Formula
Yield Approximation	$y = \dfrac{g - \frac{k}{n}}{1 + \left(\dfrac{k(n+1)}{2n}\right)}$
Bond Salesman	$y = \dfrac{g - \frac{k}{n}}{1 + 0.5k}$

3. Bond Price Items

Method	Flat Price B_{t+k}	Accrued Coupon Fr_k	Market Price = Flat Price − Accrued Coupon
Theoretical	$B_t(1+i)^k$	$Fr\left[\dfrac{(1+i)^k - 1}{i}\right]$	$B_t(1+i)^k - Fr\left[\dfrac{(1+i)^k - 1}{i}\right]$
Practical	$B_t(1+ki)$	kFr	$B_t(1+ki) - kFr$
Semi-Theoretical	$B_t(1+i)^k$	kFr	$B_t(1+i)^k - kFr$

4. Forward Price

Forward Rate $_{t1}f_{t2} = \dfrac{(1+r_2)^{t_2}}{(1+r_1)^{t_1}}$

Cost of Carry Forward Price $F = (S+s)e^{(r-c)t}$

5. Method of Equated Time

Given cash flows CF at times (t) and Equivalent Payment EP:

$$\bar{t} = \frac{\Sigma CF_t \times t}{\Sigma CF_t}$$

$$v = \frac{1}{1+i}$$

$$EPv^t = \Sigma CF_t \times v^t$$

6. Macaulay Duration and Volatility

Given cash flows CF at times (t) and discount rate (i):

$$\bar{d} = \frac{\Sigma CF_t \times v^t \times t_n}{\Sigma CF_t \times v^t}$$

$$v = \frac{\bar{d}}{1+i}$$

10. Stocks and Options

1. Capital Asset Pricing Model

$$E[r_k] = r_f + \beta_k(E[r_p] - r_f)$$

- r_k = yield rate on a security k
- r_f = Risk Free rate of interest
- r_f = yield rate on the market portfolio

- β_k = risk for security k

- $E[r_p] - r_f$ = risk premium for the market portfolio

- $\beta_k = \dfrac{\text{cov}[r_k, r_p]}{\text{var}[r_p]}$

2. Short Sale Yield

$$\text{Short Sale Yield} \quad \frac{\text{Sale Price} - \text{Buyback Price} + \text{Intereston Margin} - \text{Dividends}}{\text{Margin Requirement}}$$

3. Dividend Discount Model

The present value (PV) of dividends (D) which represent the price (P) which increase by (k) percent with a yield rate of (i) percent is:

Dividend Discount Model $PV = \dfrac{D}{i - k}$

where D = Initial Dividend Amount, k = dividend increase rate, and i = yield rate

4. Black-Scholes

Black-Scholes Formula
Call Option Value $C = S * \Phi(d_1) - E * e^{-rt} * \Phi(d_2)$
Put Option Value $P = E * e^{-rt} * (1 - \Phi(d_2)) - S(1 - \Phi(d_1))$

$$d_1 = \frac{\text{Ln}(S / E) + (r + 0.5\sigma^2) \times t}{\sigma\sqrt{t}}$$

$$d_2 = d_1 - \sigma\sqrt{t}$$

5. Options

Call Options are the right to buy a stock at a certain price with intrinsic value $S - E$

Put Options are the right to sell a stock at a certain price with intrinsic value $E - S$

Put-Call Parity $C + Ke^{-rT} = P + S_0$

Time Value = Option Value − Intrinsic Value

Calls and Puts

$$pi = \frac{\text{HighPrice} - S}{\text{HighPrice}}$$

$$pl = \frac{\text{LowPrice} - S}{\text{LowPrice}}$$

$$p = \frac{r_f - pl}{pi - pl}$$

$$C = \frac{p \times (p^+ - E)}{1 + r_f}$$

$$P = \frac{p \times (E - p^-)}{1 + r_f}$$

$$\Delta = \frac{E - p^-}{p^+ - p^-}$$

Cox Ross Rubenstein

$$C = \frac{n! \times p^k (1 - p)^{n-k} \times \max(0, u^k d^{n-k} S - E)}{rr^n (k!)(n-k)!}$$

$$P = \frac{n! \times p^k (1 - p)^{n-k} \times \max(0, E - u^k d^{n-k} S)}{rr^n (k!)(n-k)!}$$

Treynor Ratio $\Rightarrow TR = \dfrac{R_f + \beta(R_m - R_f)}{\beta}$

Weighted Average Cost of Capital $WACC = r_D \times (1 - T)D + r_E \times E$

Volatility Returns $= \dfrac{\text{Ln}(R_1)}{\text{Ln}(R_0)}$

6. Fibonacci Retracements levels

Price 1 $= P_1$ and Price 2 $= P_2$
High Price (HP) $= \text{Max}(P_1, P_2)$
Impulse Length (IL) $= P_1 - P_2$
Retracement Percentages (rp) are 23.6%, 38.2%, 50%, 61.8%, 100%
Retracements Level $= HP - IL \times rp$

11. T-Bills

T-Bill price is $P = F - \dfrac{F \times (7 \times w \times y)}{360}$ where F = Face Value, w = weeks, and y = yield percent

12. High Low Method

High Cost $= c_h$, Low Cost $= c_l$, Production High $= p_h$, Production Low $= p_l$
Variable Cost per unit (b) $= \dfrac{c_h - c_l}{p_h - p_l}$

Total Fixed Cost (TFC) $= c_h b(p_h)$
Total Fixed Cost (TFC) $= c_l b(p_l)$
Cost-Volume Formula (y): $y = TFC + bx$

DISCRETE MATH

Discrete Math Calculators

1. Truth Tables

Negation Truth Table

P	$\neg P$
T	F
F	T

Logical Disjunction Truth Table

P	Q	P ∨ Q
T	T	T
T	F	T
F	T	T
F	F	F

Logical Conjunction Truth Table

P	Q	P ∧ Q
T	T	T
T	F	F
F	T	F
F	F	F

Material Implication Truth Table

P	Q	P ⇒ Q
T	T	T
T	F	F
F	T	T
F	F	T

Material Equivalence Truth Table

P	Q	P ⇔ Q
T	T	T
T	F	F
F	T	F
F	F	T

2. Modulus Operations

Modulus Operations $\Rightarrow a \bmod b$ is the remainder left over after a is divided by b

The Quotient Remainder Theorem states that for integer n and positive integer d, $q = \dfrac{n}{d}$ and $r = n\%d$ then:

$n = dq + r$ where $0 \le r < d$

Congruence Modulo N is determined if $a \equiv b \bmod n$, then $n \mid a - b$

3. Multifactorials and Subfactorials

Multifactorial $n!^m = n \times (n-m) \times \cdots \times 1$

Subfactorial or Derangement $!n = \left[\dfrac{n!}{e} + 0.5\right]$

4. Partitions

Unordered Partitions $k = \dfrac{n}{m}$ and $a = \dfrac{n!}{(k!)(m!^k)}$

Ordered Partitions $k = \dfrac{n}{m}$ and $a = \dfrac{n!}{m!^k}$

Cross Partitions

Cross Partitions are the intersection of all items of a partition with subsets of other partitions.

$[\{1,2,3\},\{4,5,6\}][\{1,4\},\{2,3,5,6\}][\{2,8\}\{4,7,9\},\{3,6\}]$
$\{1\},\{2,3\},\{2\},\{3\},\{4\},\{5,6\},\{4\},\{6\},\{4\},\{2\},\{3,6\}$

5. Partitioned Intervals

1. Form each subinterval $S_n = [x_n, x_{n+1}]$
2. Calculate each subinterval length $\Delta_n = x_{n+1} - x_n$
3. Calculate the norm (mesh) which is the maximum value of all subinterval lengths

6. Interpolation

Linear Interpolation formula: $y_a + \dfrac{(y_b - y_a)(x - x_a)}{x_b - x_a}$

Nearest Neighbor Interpolation formula for a point p: $\sum_{k=0}^{n} |x_k - p|$, and then find the lowest difference d and use $f(d)$

7. Unique Word Arrangements

A word with M letters in it has the following unique word arrangements:

$$\frac{M}{N_1 N_2 \cdots N_M}$$

where N_i is the number of duplicates of a particular letter

8. Primitive Root

If p is prime, then b is a primitive root for p if the powers of b include all of the residue classes mod p. Use $b = 3$ and $p = 7$ as an example

n	$n - 1$	b^{n-1}	$b^{n-1} \bmod p$
1	0	$3^0 = 1$	$3^0 \bmod 7 = 1$
2	1	$3^1 = 3$	$3^1 \bmod 7 = 3$
3	2	$3^2 = 9$	$3^2 \bmod 7 = 2$
4	3	$3^3 = 27$	$3^3 \bmod 7 = 6$
5	4	$3^4 = 81$	$3^4 \bmod 7 = 4$
6	5	$3^5 = 243$	$3^5 \bmod 7 = 5$

Since we achieved all values from 1 to 6 in our residue results, then 3 is a primitive root of 7.

9. Linear Congruence

$$ax \equiv c \;(\text{mod } b)$$

1. Determine the GCF(a, b)
2. If the GCF is 1, then there will be 1 solution mod c
3. Set up the Diophantine Equation $ax + by = c$
4. Solve the Linear Congruence using Euclid Algorithm

10. Euclids Algorithm and Diophantine Equations

Diophantine Equation $ax + by = c$, then $d = gcd(a,b)$

Algorithm 35.10.1 Euclidean Algorithm
1: **procedure** FACTORIAL(a, b)
2:     $r := a \bmod b$
3:     **while** $b \neq 0$ **do**
4:         $a := b$
5:         $b := r$
6:         $r := a \bmod b$
7:     **end while**
8:     **return** a ▷ The *gcd* is the positive integer a
9: **end procedure**

Algorithm 35.10.2 Extended Euclidean Algorithm
1: **procedure** EXTEUCLID(A, B)
2:     $a := A, b := b, s := 1, t := 0, u := 0, v := 1$
3:     **while** $b \neq 0$ **do**
4:         $r := a \bmod b, q := a \div b$
5:         $a := b, b := r$
6:         $newu := s - uq, newv := t - vq$
7:         $s := u, t := v$
8:         $u := newu, v := newv$
9:     **end while**
10:    **return** a ▷ The *gcd* is the positive integer a
11: **end procedure**

Algorithm 35.11.1 Collatz Conjecture Algorithm
1: **procedure** COLLATZ(N)
2:   $n := N$
3:   **while** $n \neq 1$ **do**
4:     **if** $i \bmod 2 = 0$ **then**
5:       $n := \dfrac{n}{2}$ ▷ Even number
6:     **else**
7:       $n := 3n - 1$ ▷ Odd number
8:     **end if**
9:   **end while**
10:   **return** n ▷ Positive Integer n
11: **end procedure**

11. Collatz Conjecture

12. Prime Number Algorithms

The totient of a number n is denoted as φ

1. List all factors for n
2. for $k = 1$, $k++$, while $k < n$ List Factors for k
3. Pick up all numbers who share no factors with n other than 1

The Sieve of Eratosthenes is used to find prime numbers before a number n

1. Find the first number m less than n where $m^2 > n$
2. Subtract 1 from m: $m = m - 1$
3. List all numbers from 2 to n and call this list L
4. for $k = 2$, $k++$, while $k <= m$ Remove all multiples of k, modify L.
5. When m = k, after List is modified, the List will be $\pi(n) = L$

Fermats Little Theorem states that if p is a prime number, than for any integer a, $ap - a$ is an integer multiple of p

$$ap - 1 \equiv 1(\bmod p)$$

13. Other Number Algorithms

The Lagrange Four Square Theorem states that any natural number can be expressed as:

$$p = a^2 + b^2 + c^2 + d^2$$

14. Number Properties

Number Type	Description/Formula	First Four Examples
Perfect	Sum of Divisors = Number	6, 28, 496, 8128
Abundant	Sum of Divisors > Number	12, 18, 20, 24
Deficient	Sum of Divisors < Number	1, 2, 3, 4
Evil	Even Number of One's in Binary Expansion	3, 5, 6, 9
Odious	Odd Number of One's in Binary Expansion	1, 2, 4, 7
Triangular	each row has 1 more item in a triangular form	1, 3, 6, 10
Automorphic (Curious)	decimal expansion of n^2 ends with n	1, 5, 6, 25
Undulating	alternating digits in the form $abab$	101, 111, 121, 131,
Square	has an integer square root	1, 4, 9, 16
Cube	has an integer cube root	1, 8, 27, 64
Square	has an integer square root	1, 4, 9, 16
Palindrome	read the same forward and backward	11, 22, 33, 44
Palindromic Prime	Palindrome and Prime	101, 131, 151, 181
Repunit	Every digit equal to 1	11, 111, 1111, 11111

Apocalyptic Power	2^n contains 666	157, 192, 218, 220
Pentagonal	satisfies $\dfrac{n(3n-1)}{2}$	1, 5, 12, 22
Tetrahedral (Pyramidal)	satisfies $\dfrac{n(n+1)(n+2)}{6}$	1, 4, 10, 20
Narcissistic (Plus Perfect)	m digit number equal to the square sum of it's m-th powers of its digits	1, 2, 3, 4
Catalan	$C_n = \dfrac{2n!}{n!(n+1)!}$	1, 2, 5, 14,

LINEAR ALGEBRA

LINEAR ALGEBRA Calculators

1. Matrix Operations

Structure is a_{ij} where i = row and j = column

$$A = \begin{vmatrix} a_{11} & a_{12} & a_{13} \\ a_{21} & a_{22} & a_{23} \\ a_{31} & a_{32} & a_{33} \end{vmatrix}$$

Scalar Multiplication cA

$$cA = \begin{vmatrix} ca_{11} & ca_{12} & ca_{13} \\ ca_{21} & ca_{22} & ca_{23} \\ ca_{31} & ca_{32} & ca_{33} \end{vmatrix}$$

Matrix Addition – Add respective entries of A and B: $|A| + |B|$

$$\begin{vmatrix} a_{11} & a_{12} & a_{13} \\ a_{21} & a_{22} & a_{23} \\ a_{31} & a_{32} & a_{33} \end{vmatrix} + \begin{vmatrix} b_{11} & b_{12} & b_{13} \\ b_{21} & b_{22} & b_{23} \\ b_{31} & b_{32} & b_{33} \end{vmatrix} = \begin{vmatrix} a_{11}+b_{11} & a_{12}+b_{12} & a_{13}+b_{13} \\ a_{21}+b_{21} & a_{22}+b_{22} & a_{23}+b_{23} \\ a_{31}+b_{31} & a_{32}+b_{32} & a_{33}+b_{33} \end{vmatrix}$$

Matrix Subtraction – Subtract respective entries of A and B: $|A| - |B|$

$$\begin{vmatrix} a_{11} & a_{12} & a_{13} \\ a_{21} & a_{22} & a_{23} \\ a_{31} & a_{32} & a_{33} \end{vmatrix} - \begin{vmatrix} b_{11} & b_{12} & b_{13} \\ b_{21} & b_{22} & b_{23} \\ b_{31} & b_{32} & b_{33} \end{vmatrix} = \begin{vmatrix} a_{11} - b_{11} & a_{12} - b_{12} & a_{13} - b_{13} \\ a_{21} - b_{21} & a_{22} - b_{22} & a_{23} - b_{23} \\ a_{31} - b_{31} & a_{32} - b_{32} & a_{33} - b_{33} \end{vmatrix}$$

Matrix Multiplication: $|A| \times |B|$

$$\begin{vmatrix} a_{11} & a_{12} & a_{13} \\ a_{21} & a_{22} & a_{23} \\ a_{31} & a_{32} & a_{33} \end{vmatrix} \times \begin{vmatrix} b_{11} & b_{12} & b_{13} \\ b_{21} & b_{22} & b_{23} \\ b_{31} & b_{32} & b_{33} \end{vmatrix} =$$

$$\begin{vmatrix} a_{11}b_{11} + a_{12}b_{21} + a_{13}b_{31} & a_{11}b_{12} + a_{12}b_{22} + a_{13}b_{32} & a_{11}b_{13} + a_{12}b_{23} + a_{13}b_{33} \\ a_{21}b_{11} + a_{22}b_{21} + a_{23}b_{31} & a_{21}b_{12} + a_{22}b_{22} + a_{23}b_{32} & a_{21}b_{13} + a_{22}b_{23} + a_{23}b_{33} \\ a_{31}b_{11} + a_{32}b_{21} + a_{33}b_{31} & a_{31}b_{12} + a_{32}b_{22} + a_{33}b_{32} & a_{31}b_{13} + a_{32}b_{23} + a_{33}b_{33} \end{vmatrix}$$

2. Matrix Properties

Determinant
Inverse
Adjoint

3. Cross Product in R3

$$\begin{vmatrix} i & j & k \\ x_1 & x_2 & x_3 \\ y_1 & y_2 & y_3 \end{vmatrix}$$

$$X \cdot Y = (x_2 y_3 - x_3 y_2)i + (x_3 y_1 - x_1 y_3)j + (x_1 y_2 - x_2 y_1)k$$

$$A = \begin{vmatrix} 1 & y_1 & z_1 \\ 1 & y_2 & z_2 \\ 1 & y_3 & z_3 \end{vmatrix} \quad B = \begin{vmatrix} x_1 & 1 & z_1 \\ x_2 & 1 & z_2 \\ x_3 & 1 & z_3 \end{vmatrix} \quad C = \begin{vmatrix} x_1 & y_1 & 1 \\ x_2 & y_2 & 1 \\ x_3 & z_3 & 1 \end{vmatrix}$$

$$D = \begin{vmatrix} x_1 & y_1 & z_1 \\ x_2 & y_2 & z_2 \\ x_3 & z_3 & z_3 \end{vmatrix}$$

$$|A| = y_1(z_2 - z_3) + y_2(z_3 - z_1) + y_3(z_1 - z_2)$$

$$|B| = z_1(x_2 - x_3) + z_2(x_3 - x_1) + z_3(x_1 - x_2)$$

$$|C| = x_1(y_2 - y_3) + x_2(y_3 - y_1) + x_3(y_1 - y_2)$$

$$|D| = x_1(y_2 z_3 - y_3 z_2) + x_2(y_3 z_1 - y_1 z_3) + x_3(y_1 z_2 - y_2 z_1)$$

4. Digraph

Digraph

5. Equation of a Plane

The standard equation for a plane is: $Ax + By + Cz + D = 0$
Given three points in space $(x_1, y_1, z_1), (x_2, y_2, z_2), (x_3, y_3, z_3)$, the equation of the plane through these points is given by the following determinants:

6. Plane and Parametric Equations in R3

Given a vector $A = \begin{vmatrix} a \\ b \\ c \end{vmatrix}$ and a point (x_0, y_0, z_0)

The plane equation passing through (x_0, y_0, z_0) and perpendicular to A is denoted as

$$a(x - x_0) + b(y - y_0) + c(z - z_0) = 0$$

This is a parametric equation of L, which can be written in terms of the components as:

$$x = x_0 + tu$$
$$y = y_0 + tv \ (-\infty < t < \infty)$$
$$z = z_0 + tw$$

7. Vectors

Given two vectors $A = \begin{vmatrix} a_1 \\ a_2 \\ a_3 \end{vmatrix} \quad B = \begin{vmatrix} b_1 \\ b_2 \\ b_3 \end{vmatrix}$

Length (Magnitude) $\| A \| = \sqrt{a_1^2 + a_2^2 + a_3^2}$

$$A + B = (a_1 + b_1, a_2 + b_2, a_3 + b_3)$$
$$A - B = (a_1 - b_1, a_2 - b_2, a_3 - b_3)$$
$$A \cdot B = a_1 b_1 + a_2 b_2 + a_3 b_3$$

Distance: $\overline{AB} = \sqrt{(a_1 - b_1)^2 + (a_2 - b_2)^2 + (a_3 - b_3)^2}$

For $\angle\theta$ between A and B, $\cos(\theta) = \dfrac{\| A \| \cdot \| B \|}{\| A \| \times \| B \|}$

Unit Vector $\dfrac{U}{\| A \|}$

Vectors are perpendicular if $| A | \cdot | B | = 0$
Vectors are parallel if $| A | \cdot | B | = \| A \| \times \| B \|$
Cauchy-Schwartz $| A \cdot B | \leq \| A \| \times \| B \|$

Orthogonal Projection $\text{proj}_B A = \dfrac{(A \cdot B) \times B}{\| B \|^2}$

8. Markov Chain

The Markov Chain procedure with Transition Matrix T and initial state vector $P^0 \Rightarrow P^{(n)} = TP^{(n-1)}$

SET THEORY

Set Theory Calculators

1. Roster Notation Items

Roster Notation	Symbol	Action
Greater than	>	Do not include the last number on the right
Greater than or equal to	≥	Include the last number on the right
Less than	<	Do not include the last number on the left
Less than or equal to	≤	Include the last number on the left

The set of all natural numbers less than 8: $\{1, 2, 3, 4, 5, 6, 7\}$

The set of all natural numbers less than or equal to 8: $\{1, 2, 3, 4, 5, 6, 7, 8\}$

The set of all integers less than 5: $\{-\infty, ..., 0, 1, 2, 3, 4\}$

The set of all integers greater than or equal to 5: $\{5, 6, 7, ..., \infty\}$

2. Set Notation Items

Set Notation Items	Formula
Union (All Elements in A and B)	$A \cup B$
Intersection (All Elements in A and B	$A \cap B$
Set Difference for A (All Elements in A not in B	$A - A \cap B$
Set Difference for B (All Elements in B not in A	$B - A \cap B$
Symmetric Difference (All Elements in A not in B plus All elements in B not in A	$A \Delta B = (A - B) \cup (B - A)$
Concatenation (All Elements in A and B)	$A \cdot B$
Cardinality (Number of Elements in A)	$|A|$
Cardinality (Number of Elements in B)	$|B|$
Cartesian Product (All Orderd Pairs in A and B)	AxB
Power Set (All Subsets of A including)	$P(A)$
Jaccard Index $J(A,B)$	$\dfrac{|A \cap B|}{|A \cup B|}$
Jaccard Distance $J_\sigma(A,B)$	$1 - J(A,B)$
Dice Coefficient s	$\dfrac{|A \cap B|}{|A| + |B|}$

3. Power Sets

Power Set is the set of all subsets including the empty set.
$$S = \{a, b, c, d\}$$
When S contains n items, the Power Set P should contain 2^n items

The power set
$$P = \{\varnothing, a, b, c, d, (a, b), (a, c), (a, d), (b, c), (b, d), (c, d), (a, b, c), (a, b, d), (a, c, d),$$
$$(b, c, d), (a, b, c, d)\}$$

4. Partitions of a Set

The number of partitions of a set is the Bell Number $B_{n+1} = \sum_{k=0}^{n} \frac{n!}{k!(n-k)!} B_k$

CHAPTER 39

ECONOMICS

ECONOMICS Calculators

1. Formulas

Budget Line Equation	$I = Q_x P_x + Q_y P_y$
Net Exports (NX)	$NX = E - I$
Gross Domestic Product	$GDP = C + I + G + (E - I)$
Gross Domestic Product Deflator	$GDP\ \text{Deflator} = \dfrac{GDP_{\text{nominal}} \times 100}{GDP_{\text{real}}}$
Equation of Exchange	$MV = PQ$
Money Multiplier	$\dfrac{1}{RR}$
Gross Profit	$\text{GrossProfit} = \text{Revenue} - \text{Cost}$
Gross Profit Margin	$\dfrac{\text{Gros Profit}}{\text{Revenue}}$
Net Profit	$\text{Net Profit} = \text{Gross Profit}(1 - \text{Tax Rate})$
Net Profit Margin	$\dfrac{\text{NetProfit}}{\text{Revenue}}$
Sectoral Balance	$PS - PI = GS - T + E - I$

Cost Utility Ratio	$\dfrac{\text{Cost}}{\text{Utility}}$
Total Revenue (TR)	$TR = \text{Cost} \times \text{Quantity}$
Herfindahl Index (H)	Σs_i^2
Normalized Herfindahl Index H*	$H^* = \dfrac{H - 1/N}{1 - 1/N}$

CHEMISTRY

1. Laws

P = Pressure, V = Volume, T = Temperature

Boyle's Law $\Rightarrow P_1V_1 = P_2V_2$

Charles Law $\Rightarrow V_1T_2 = V_2T_1$

Combined Gas Law $\Rightarrow P_1V_1T_2 = P_2V_2T_1$

Pressure Law $\Rightarrow P_1T_2 = P_2T_1$

2. Density

Density $\Rightarrow D = MV$ where M = Mass and V = Volume

BIOLOGY

1. Punnett Square Items

1. Punnett Square

Punnett Square where A = Dominant Gene and a = Recessive Gene

	A	a
A	AA	Aa
a	Aa	aa

2. Hardy Weinberg

Hardy Weinberg where A = Dominant Gene and a = Recessive Gene with probabilities p and $q = 1 - p$

	A (p)	a (q)
A (p)	AA (p^2)	Aa (pq)
a (q)	Aa (pq)	aa (q^2)

ACTUARIAL SCIENCE

1. Population and Mortality Factors

Population Formula	$d_x = l_x - l_{x+1}$
Survival Formula	$p_x = \dfrac{l_{x+1}}{l_x}$
Mortality Formula	$q_x = 1 - p_x$

2. Hazard Rate Function

Also called the force of mortality: $\lambda(x) = \dfrac{f_X(x)}{S_X(x)}$
Density divided by Survival

3. Social Security and Covered Compensation

Social Security Calculation
Averaged Indexed Monthly Earnings

$$AIME = \frac{35 \text{ year Salary History with Indexing}}{35 \times 12}$$

$\text{Tier1} = 0.9 \times \min(\text{AIME}, \text{Bend Point1})$
$\text{Tier2} = 0.32 \times \text{Maximum}(\text{Minimum}(\text{Bend Point2}, \text{AIME}) - \text{Bend Point1},$
$\text{Tier3} = 0.15 \times \text{Max}(0, \text{AIME} - \text{Bend Point2})$
Primary Social Security Benefit $PSSB = Tier1 + Tier2 + Tier3$
Reduction Factor 1:

$$\frac{5}{9} \times 0.01 \times \text{Min}\,(36, \text{Months Retiring before } SSNRA)$$

Reduction Factor 2:

$$\frac{5}{9} \times 0.01 \times \text{Maximum}\,(\text{Months Retiring before } SSNRA - 36, 0)$$

Early Reduction Factor: $ERF = RED1 + RED2$
Reduced Social Security if taken before Primary Social Security Age:
$PSSB \times ERF$
Covered Compensation

- Take birth year + Social Security Normal Retirement Age
- Add up the 35 years of compensation history
- Divide this by 35 to get your covered compensation

ENGINEERING

ENGINEERING Calculators

1. Bending Beams Displacement

Bending Beams Displacement (dual Support) $\Rightarrow d = \dfrac{FI^3}{4\,yab^3}$

Bending Beams Displacement (Single Support) $\Rightarrow d = \dfrac{FI^3}{yab^3}$

2. Direct Current Ohms Law

Formulas

$P = V * I$

$R = \dfrac{V}{I}$

$P = I^2 R$

$I = \dfrac{V}{R}$

$$V = \sqrt{P} \times R$$

I = Current (Amps), V = (volts), R = Resistance (ohms), P = Power (watts)

3. Sine Wave

Formulas
Average Value $= 0.637 \times$ Peak Value
$RMS = 0.707 \times$ Peak Value
$RMS = 1.11 \times$ Average Value
Peak Value $= 1.57 \times$ Average Value
Peak Value $= 1.414 \times RMS$
Average $= 0.9 \times RMS$

4. Young's Modulus

Youngs Modulus $\Rightarrow Y = \dfrac{n}{m^2}$

5. Static Determinancy and Stability

Static Determinancy and Stability $\Rightarrow 2j = m + 3$

CAPITAL BUDGETING

Calculators

1. Average Returns

$$\text{Average Return} = \frac{\sum_{i=1}^{n} CF}{n}$$

$$\text{Average Rate of Return} = \frac{AR}{I}$$

2. Incremental Cash Flow

ICF = Cash Inflows Cash Outflows (Inflows Outflows Depreciation Expense) × TaxRate

Taxes = (Inflows Outflows Depreciation Expense) × Tax Rate

3. Net Present Value and IRR and Profitability Index

$$PV_t = \frac{CF_t}{(1+i)^t}$$

$$NPV = NPV = \sum_{t=0}^{n} PV_t$$

4. Modified Internal Rate of Return (MIRR)

$$MIRR = MIRR = \sqrt[n]{\frac{FV \text{ of Positive} CF}{PV \text{ of Negative} CF}}$$

5. Equivalent Annual Cost

$$EAC = \text{Discounted Investment} + \text{Maintenance Cost}$$

PHYSICS

PHYSICS

$$\text{Acceleration} = \frac{v - v_0}{t}$$

$$\text{Angular Momentum} = L = M \times V \times R$$

$$\text{Centripetal Acceleration} = \frac{v^2}{r}$$

$$\text{Frequency and Wavelength } f = \frac{c}{\lambda}$$

$$\text{Gravitational Force } F = \frac{G \times m1 \times m2}{d^2}$$

Kinematic Equations

$$d = vi_t + \frac{1}{2}at_2$$

$$v_f^2 = v_i^2 + 2ad$$

$$v_f = v_i + at$$

$$d = \frac{1}{2}(v_i + v_f)t$$

Kinetic Energy $KE = \frac{1}{2}MV^2$

Lever Systems Formula $F_1 x = F_2(d - x)$

Little's Law $WIP = CT * TH$

Moment of Inertia $= \text{Mass} \times \text{Length}^2$

Parallel Resistors $\dfrac{1}{\dfrac{1}{R_1} + \dfrac{1}{R_2} + ... + \dfrac{1}{R_n}}$

Work $= F \times D$

ACCOUNTING

ACCOUNTING Calculators

1. Balance Sheet

Working Capital = Current Assets – Current Liabilities

$$\text{Current Ratio} = \frac{\text{Current Assets}}{\text{Current Liabilities}}$$

Quick Assets = Current Assets – Inventory

$$\text{Quick / Acid Test / Current Ratio} = \frac{\text{Quick Assets}}{\text{Current Liabilities}}$$

Receivable Turnover

2. Methods

Installment Sales Method of Accounting

$$\frac{\text{Quick Assets}}{\text{Current Liabilities}} \text{ and Cost Percentage} = 1 - \text{Gross Profit Percentage}$$

Chain Discount $Pd_1d_2 \cdots d_n = P(1-d_1)(1-d_2) \cdots (1-d_n)$

Percentage of Completion $\Rightarrow$ Gross Profit to Date = Percent Complete $\times$ Profit Amount

Vendor Discount a/b net c $\Rightarrow$ $\dfrac{a}{100-a} \times \dfrac{360}{c-b}$

3. Inventory

FIFO and LIFO

Average Inventory $\dfrac{I_0 + I_1}{2}$

Inventory Turnover Ratio $\dfrac{COGS}{AI}$

H E A L T H   A N D   W E L L
B E I N G

Health Calculators

Basal Metabolic Rate	$BMR = 66 + 13.7x$ weight in pounds $+ 5x$ height in inches -6.8 age in years
Body Mass Index	$BMI = \dfrac{\text{Weight in Kilos}}{\text{Height In Meters}^2}$
Cholesterol	Cholesterol $= HDL + LDL + \dfrac{1}{5}$ Triglycerides

CHAPTER 48

WORD PROBLEMS

Word Problem Calculators

2 number word problems 2 numbers a and b have a sum of s and a product p.

$a + b = s$ and $ab = p$

Consecutive Integer Word Problems 2 consecutive integers have a sum of s.

Number 1 $= n$, Number 2 $= n + 1$, so $n + (n + 1) = s$

Distance Problems $d = rt$

Inclusive Number Word Problem Type	Formula
Average of all inclusive numbers A to B	$\dfrac{A + B}{2}$
Count of all inclusive numbers A to B	$B - A + 1$
Sum of all inclusive numbers A to B	$\dfrac{(A + B)(B - A + 1)}{2}$

Number Type	Sum of the First Formula	First Five Numbers
Whole Numbers	$S_n = \dfrac{n(n-1)}{2}$	$0, 1, 2, 3, 4$
Natural Numbers	$S_n = \dfrac{n(n+1)}{2}$	$1, 2, 3, 4, 5$

Even Numbers	$S_n = n(n+1)$	2, 4, 6, 8, 10
Odd Numbers	$S_n = n^2$	1, 3, 5, 7, 9
Square Numbers	$S_n = \dfrac{n(n+1)(2n+1)}{6}$	1, 4, 9, 16, 25
Cube Numbers	$S_n = \dfrac{n^2(n+1)^2}{4}$	1, 8, 27, 64, 125
Fourth Power Numbers	$S_n = \dfrac{n(n+1)(2n+1)(3n^2+3n-1)}{30}$	1, 16, 81, 256, 625

CONVERSIONS

1. Coin Conversions

1 dollar	100 pennies
1 dollar	20 nickels
1 dollar	10 dimes
1 dollar	4 quarters
1 dollar	2 half-dollars
1 half-dollar	50 pennies
1 half-dollar	10 nickels
1 half-dollar	5 dimes
1 half-dollar	2 quarters
1 quarter	25 pennies
1 quarter	5 nickels
1 dime	10 pennies
1 dime	2 nickels
1 nickel	5 pennies

2. Temperature Conversions

$F = (1.8C) + 32$

$K = (C + 273.15)$

$R = \dfrac{9(C + 273.15)}{5}$

$N = 0.3333C$

$Reaumur = 0.8C$

$C = \dfrac{9}{5}(F - 32)$

$K = \dfrac{5}{9}(F + 459.67)$

$R = F + 459.67$

$N = \dfrac{11}{60}(F32)$

$R = \dfrac{4}{9}(F32)$

$C = K273.15$

$F = \dfrac{9}{5}K - 459.67$

$R = \dfrac{9}{5}K$

$N = 0.33(F273.15)$

$Reaumur = 0.8(K273.15)$

3. Roman Numerals

Roman Numeral Letter	Number Value
I	1
II	2
III	3
IV	4
V	5
X	10
L	50
C	100
D	500
M	1,000

To convert a number to Roman Numeral notation, use these steps:

1. Find the highest number in the table less than or equal to your number.
2. Start with that letter
3. Take your number and subtract the table number This is your new number
4. Repeat steps 1-3 until you reach zero as your target number

In Excel, you can use the ROMAN function to convert numbers to Roman Numerals. Example: =ROMAN(50)

4. ROT 13

Original Letter	Alphabet Position	Alphabet Position Mod 13	ROT 13 Letter
A	1	1 + 13 = 14	N
B	2	2 + 13 = 15	O
C	3	3 + 13 = 16	P

D	4	4 + 13 = 17	Q
E	5	5 + 13 = 18	R
F	6	6 + 13 = 19	S
G	7	7 + 13 = 20	T
H	8	8 + 13 = 21	U
I	9	9 + 13 = 22	V
J	10	10 + 13 = 23	W
K	11	11 + 13 = 24	X
L	12	12 + 13 = 25	Y
M	13	13 + 13 = 26	Z
N	14	14 + 13 = 27 (Reset portion greater than 26 27 − 26 = 1)	A
O	15	15 + 13 = 28 (Reset portion greater than 26 28 − 26 = 2)	B
P	16	16 + 13 = 29 (Reset portion greater than 26 29 − 26 = 3)	C
Q	17	17 + 13 = 30 (Reset portion greater than 26 30 − 26 = 4)	D
R	18	18 + 13 = 31 (Reset portion greater than 26 31 − 26 = 5)	E
S	19	19 + 13 = 32 (Reset portion greater than 26 32 − 26 = 6)	F
T	20	20 + 13 = 33 (Reset portion greater than 26 33 − 26 = 7)	G
U	21	21 + 13 = 34 (Reset portion greater than 26 34 − 26 = 8)	H
V	22	22 + 13 = 35 (Reset portion greater than 26 35 − 26 = 9)	I
W	23	23 + 13 = 36 (Reset portion greater than 26 36 − 26 = 10)	J
X	24	24 + 13 = 37 (Reset portion greater than 26 37 − 26 = 11)	K
Y	25	25 + 13 = 38 (Reset portion greater than 26 38 − 26 = 12)	L
Z	26	26 + 13 = 39 (Reset portion greater than 26 39 − 26 = 13)	M

Algorithm 48.4.1 Rot13 Algorithm

1: **procedure** ROT13(*word*)

2:    *nlen* = *word.length*

3:    *phrase* = *null*

4:    **for** *i*:= 1, *i* ≤ *n*, *i* + + **do**   ▷ Start at the first letter and
                                                                      go to the end

5:        *letter* = *n* mod 26

6:        *phrase* := *phrase* + *LetterNumber*(*letter*)

7:    **end for**

8:    **return** *phrase*   ▷ Return the numeric phrase

9: **end procedure**

---

5. Area Conversions

Letter	Number Value
1 hectare	2.4711 acres
1 square mile	640 acres
1 square kilometer	247.105 acres
1 square mile	258.998811 hectares
1 square kilometer	100 hectares
1 acre	6272640 square inchs
1 hectare	15500031 square inchs
1 square foot	144 square inchs
1 square yard	1296 square inchs
1 square mile	4014489600 square inchs
1 square meter	1550.0031 square inchs
1 square kilometer	1550003100 square inchs
1 acre	43560 square foots
1 hectare	107639.104 square foots
1 square yard	9 square foots
1 square mile	27878400 square foots
1 square meter	10.7639 square foots

1 square kilometer	10763910.4 square foots
1 acre	4840 square yards
1 hectare	11959.9005 square yards
1 square mile	3097600 square yards
1 square meter	1.19599005 square yards
1 square kilometer	1195950.05 square yards
1 acre	4046856420 square millimeters
1 hectare	10000000000 square millimeters
1 square inch	645.16 square millimeters
1 square foot	92903.04 square millimeters
1 square yard	836127.36 square millimeters
1 square mile	2589988110336 square millimeters
1 square meter	1000000 square millimeters
1 square kilometer	1000000000000 square millimeters
1 acre	4046.86 square meters
1 hectare	10000 square meters
1 square mile	2589988.11 square meters
1 square kilometer	1000000 square meters
1 square mile	2.59 square kilometers

6. Computer Storage Conversions

1 kilobyte	1024 bytes
1 megabyte	1048576 bytes
1 gigabyte	1073741824 bytes
1 terabyte	1099511627776 bytes
1 petabyte	1125899906842620 bytes
1 exabyte	1152921504606840000 bytes
1 megabyte	1024 kilobytes
1 gigabyte	1048576 kilobytes
1 terabyte	1073741824 kilobytes
1 petabyte	1099511627776 kilobytes
1 exabyte	1125899906842620 kilobytes

1 gigabyte	1024 megabytes
1 terabyte	1048576 megabytes
1 petabyte	1073741824 megabytes
1 exabyte	1099511627776 megabytes
1 terabyte	1024 gigabytes
1 petabyte	1048576 gigabytes
1 exabyte	1073741824 gigabytes
1 petabyte	1024 terabytes
1 exabyte	1048576 terabytes
1 exabyte	1024 petabytes

7. Liquid Conversions

1 ounce	29.5735296 milliliters
1 ounce	6 teaspoons
1 ounce	2 tablespoons
1 cup	8 ounces
1 cup	240 milliliters
1 cup	48 teaspoons
1 cup	16 tablespoons
1 liter	1000 milliliters
1 teaspoon	5 milliliters
1 tablespoon	15 milliliters
1 quart	946.352946 milliliters
1 gallon	3785.41178 milliliters
1 pint	473.176473 milliliters
1 gallon	3.78541178 liters
1 liter	202.884136 teaspoons
1 tablespoon	3 teaspoons
1 quart	192 teaspoons
1 gallon	768 teaspoons
1 pint	96 teaspoons
1 liter	67.6280454 tablespoons
1 quart	64 tablespoons

1 gallon	256 tablespoons
1 pint	32 tablespoons
1 liter	4.22675284 cups
1 quart	4 cups
1 gallon	16 cups
1 pint	2 cups
1 liter	1.05668821 quarts
1 gallon	4 quarts
1 liter	2.11337642 pints
1 quart	2 pints
1 gallon	8 pints
1 pint	16 ounces
1 liter	33.8140227 ounces
1 quart	32 ounces
1 gallon	128 ounces
1 bushel	1191.57478 ounces
1 bushel	148.946848 cups
1 bushel	74.4734238 pints
1 bushel	35239.072 milliliters
1 bushel	35.239072 liters
1 bushel	7149.44868 teaspoons
1 bushel	2383.14956 tablespoons
1 bushel	37.2367119 quarts
1 bushel	9.30917797 gallons
1 ounce	29573.5296 microliters
1 cup	236588.236 microliters
1 pint	473176.473 microliters
1 milliliter	1000 microliters
1 liter	1000000 microliters
1 deciliter	100000 microliters
1 tablespoon	14786.7648 microliters
1 teaspoon	4928.92159 microliters
1 quart	946352.946 microliters

1 gallon	3785411.78 microliters
1 bushel	35239072 microliters
1 deciliter	3.38140227 ounces
1 cup	2.36588237 deciliters
1 pint	4.73176473 deciliters
1 deciliter	100 milliliters
1 liter	10 deciliters
1 quart	9.46352946 deciliters
1 gallon	37.8541178 deciliters
1 bushel	352.39072 deciliters

8. Speed Conversions

1 mph	1.47 ft/ss
1 m/sec	3.281 ft/ss
1 mph	1.61 km/hs
1 ft/s	1.097 km/hs
1 m/sec	3.6 km/hs
1 mph	2.236 m/secs

9. Linear Conversions

1 inch	25.4 millimeters
1 inch	2.54 centimeters
1 inch	25,400 micrometers
1 foot	12 inches
1 foot	304.8 millimeters
1 foot	30.48 centimeters
1 foot	304,800 micrometers
1 yard	36 inches
1 yard	3 feet
1 yard	914.4 millimeters
1 yard	91.44 centimeters

1 yard	914,400 micrometers
1 mile	63,360 inches
1 mile	5,280 feet
1 mile	1,760 yards
1 mile	1,609,344 millimeters
1 mile	160,934.4 centimeters
1 mile	1,609.344 meters
1 mile	1.609344 kilometers
1 mile	8 furlongs
1 mile	1,609,344,000 micrometers
1 millimeter	1,000 micrometers
1 centimeter	10 millimeters
1 centimeter	10,000 micrometers
1 meter	39.3700787 inches
1 meter	3.2808399 feet
1 meter	1.0936133 yards
1 meter	1,000 millimeters
1 meter	100 centimeters
1 meter	1,000,000 micrometers
1 kilometer	39,370.0787 inches
1 kilometer	3,280.8399 feet
1 kilometer	1,093.6133 yards
1 kilometer	1,000 meters
1 kilometer	1,000,000 millimeters
1 kilometer	100,000 centimeters
1 kilometer	4.97096954 furlongs
1 kilometer	1,000,000,000 micrometers
1 furlong	7,920 inches
1 furlong	660 feet
1 furlong	220 yards
1 furlong	201,168 millimeters
1 furlong	20,116.8 centimeters
1 furlong	201.168 meters
1 furlong	201,168,000 micrometers

10. RGB HEX Conversions

RGB to HEX using (R, G, B) as a numerical entry

$R1 = \dfrac{R - R \bmod 16}{16}$ and pick off the number letter for

0123456789ABCDEF

$R2 = R \bmod 16$ and pick off the number letter for
0123456789ABCDEF

$G1 = \dfrac{G - G \bmod 16}{16}$ and pick off the number letter for

0123456789ABCDEF

$G2 = G \bmod 16$ and pick off the number letter for
0123456789ABCDEF

$B1 = \dfrac{B - B \bmod 16}{16}$ and pick off the number letter for

0123456789ABCDEF

$B2 = B \bmod 16$ and pick off the number letter for
0123456789ABCDEF

Hue $\Rightarrow h = \dfrac{60°(G - B)}{R - B}$

11. 12 and 24 Hour Clock Conversions

24 hour to 12 hour conversions: Any PM hours get switched to mod 13.

12. Time Conversions

1 second	1000 milliseconds
1 minute	60000 milliseconds
1 hour	3600000 milliseconds
1 day	86400000 milliseconds
1 week	604800000 milliseconds
1 fortnight	8467200000 milliseconds
1 month	2592000000 milliseconds
1 quarter	7776000000 milliseconds
1 year	31104000000 milliseconds
1 decade	311040000000 milliseconds
1 century	3110400000000 milliseconds
1 millenium	31104000000000 milliseconds
1 millisecond	1000000 nanoseconds
1 millisecond	1000 microseconds
1 minute	60 seconds
1 hour	3600 seconds
1 day	86400 seconds
1 week	604800 seconds
1 fortnight	8467200 seconds
1 month	2592000 seconds
1 quarter	7776000 seconds
1 year	31104000 seconds
1 decade	311040000 seconds
1 century	3110400000 seconds
1 millenium	31104000000 seconds
1 second	1000000000 nanoseconds
1 hour	60 minutes
1 day	1440 minutes
1 week	10080 minutes
1 fortnight	141120 minutes
1 month	43200 minutes
1 quarter	129600 minutes

1 year	518400 minutes
1 decade	5184000 minutes
1 century	51840000 minutes
1 millenium	518400000 minutes
1 minute	60000000000 nanoseconds
1 day	24 hours
1 week	168 hours
1 fortnight	2352 hours
1 month	720 hours
1 quarter	2160 hours
1 year	8640 hours
1 decade	86400 hours
1 century	864000 hours
1 millenium	8640000 hours
1 week	7 days
1 fortnight	98 days
1 month	30 days
1 quarter	90 days
1 year	360 days
1 decade	3600 days
1 century	36000 days
1 millenium	360000 days
1 fortnight	2 weeks
1 month	4 weeks
1 quarter	1.5 weeks
1 year	52 weeks
1 decade	60 weeks
1 century	600 weeks
1 millenium	6000 weeks
1 month	2 fortnights
1 quarter	6 fortnights
1 year	24 fortnights
1 decade	240 fortnights

1 century	2400 fortnights
1 millenium	24000 fortnights
1 quarter	3 months
1 year	12 months
1 decade	120 months
1 century	1200 months
1 millenium	12000 months
1 year	4 quarters
1 decade	40 quarters
1 century	400 quarters
1 millenium	4000 quarters
1 decade	10 years
1 century	100 years
1 millenium	1000 years
1 century	10 decades
1 microsecond	1000 nanoseconds
1 second	1000000 microseconds
1 minute	60000000 microseconds
1 hour	3600000000 microseconds
1 day	86400000000 microseconds
1 week	604800000000 microseconds
1 fortnight	1.2096E+12 microseconds
1 month	2.63E+12 microseconds
1 quarter	7.89E+12 microseconds
1 year	3.156E+13 microseconds
1 decade	3.156E+14 microseconds
1 century	3.156E+15 microseconds
1 millenium	3.156E+16 microseconds

13. Unit Conversions

1 pair	2 units
1 half-dozen	6 units
1 half-dozen	3 pairs
1 dozen	12 units
1 dozen	6 pairs
1 dozen	2 half-dozens
1 bakers-dozen	13 units
1 gross	144 units
1 gross	72 pairs
1 gross	24 half-dozens
1 gross	12 dozens

14. Weight Conversions

1 pound	16 ounces
1 kilogram	35.2739619 ounces
1 stone	224 ounces
1 ton	32000 ounces
1 kilogram	2.20462262 pounds
1 stone	14 pounds
1 ton	2000 pounds
1 ounce	28349.5231 milligrams
1 pound	453592.37 milligrams
1 gram	1000 milligrams
1 kilogram	1000000 milligrams
1 stone	6350293.8 milligrams
1 ton	907184740 milligrams
1 centigram	10 milligrams
1 ounce	28.3495231 grams
1 pound	453.59237 grams
1 kilogram	1000 grams

1 stone	6350.29318 grams
1 ton	907184.74 grams
1 stone	6.35029318 kilograms
1 ton	907.18474 kilograms
1 ton	142.857143 stones
1 ounce	2834.951826 centigrams
1 pound	45359.22922 centigrams
1 gram	100 centigrams
1 kilogram	100000 centigrams
1 stone	625000 centigrams
1 ton	90718474 centigrams
1 ounce	28350000 micrograms
1 pound	453600000 micrograms
1 milligram	1000 micrograms
1 gram	1000000 micrograms
1 kilogram	1000000000 micrograms
1 centigram	10000 micrograms
1 stone	6350000000 micrograms
1 ton	907200000000 micrograms

OTHER CALCULATIONS

Miscellaneous Calculators

1. Mathematical Constants

Constant	Value
π – Archimedes Constant	3.141592653589793
e	2.181828
Pythagoras Constant	$\sqrt{2}$
i	$\sqrt{-1}$
ϕ Golden Ratio	$\dfrac{1+\sqrt{5}}{2}$
Gamma γ	$\lim_{n \to \infty} \sum_{k=1}^{n} \frac{1}{k} - \mathrm{Ln}(n) = 0.577215664901$
Gelfonds Constant	e^{π}

2. Translators

1. Morse Code

Character	Morse Code Translation	Dit(Dah)	Dot Count	Dash Count
A	·	di-dah	1	1
B	···	dah-di-di-dit	3	1
C	··	dah-di-dah-dit	2	2
D	··	dah-di-dit	2	1
E	·	dit	1	0
F	···	di-di-dah-dit	3	1
G	·	dah-dah-dit	1	2
H	····	di-di-di-dit	4	0
I	··	di-dit	2	0
J	·	di-dah-dah-dah	1	3
K	.	dah-di-dah	1	2
L	···	di-dah-di-dit	3	1
M		dah-dah	0	2
N	·	dah-dit	1	1
O		dah-dah-dah	0	3
P	··	di-dah-dah-dit	2	2
Q	·	dah-dah-di-dah	1	3
R	··	di-dah-dit	2	1
S	···	di-di-dit	3	0
T		dah	0	1
U	··	di-di-dah	2	1
V	···	di-di-di-dah	3	1
W	·	di-dah-dah	1	2
X	·	dah-di-di-dah	2	2
Y	·	dah-di-dah-dah	1	3
Z	··	dah-dah-di-dit	2	2

0		dah-dah-dah-dah-dah	0	5
1	.	di-dah-dah-dah-dah	1	4
2	..	di-di-dah-dah-dah	2	3
3	...	di-di-di-dah-dah	3	2
4	....	di-di-di-di-dah	4	1
5	.....	di-di-di-di-dit	5	0
6	....	dah-di-di-di-dit	4	1
7	...	dah-dah-di-di-dit	3	2
8	..	dah-dah-dah-di-dit	2	3
9	.	dah-dah-dah-dah-dit	1	4

2. Phone Number

Phone Number Translation

$A, B, C \Rightarrow 2$
$D, E, F \Rightarrow 3$
$G, H, I \Rightarrow 4$
$J, K, L \Rightarrow 5$
$M, N, O \Rightarrow 6$
$P, Q, R, S \Rightarrow 7$
$T, U, V \Rightarrow 8$
$W, X, Y, Z \Rightarrow 9$

3. Affine Cipher

Affine Cipher

$$E(x) = (ax + b) \bmod m \text{ and } D(x) = a^{-1}(x - b) \bmod m$$

3. Bitwise Operations and Shifting

2 binary numbers a and b with digit place d

Bitwise AND a_d and b_d must both be 1

Bitwise OR either a_d or b_d must be 1

Bitwise XOR either $a_d = 1$ and $b_d = 0$ or $a_d = 0$ and $b_d = 1$

Bitwise NOT digits are reversed: $a_d = 1, \Rightarrow a_d = 0$ and $a_d = 0, \Rightarrow a_d = 1$

Bit Shift Left $a << b = a * 2^b$

Bit Shift Right $a >> b = \frac{a}{2^b}$

GAMING

1. Dice

1 Die Roll	Probability
Any Number	$\frac{1}{6}$
Even (2,4,6)	$\frac{1}{2}$
Odd (1,3,5)	$\frac{1}{2}$
Prime (1,2,3,5)	$\frac{2}{3}$
Composite (4,6)	$\frac{1}{3}$

2 Die Roll	Die Combos	Probability
Sum = 2	(1,1)	$\frac{1}{36}$
Sum = 3	(1,2), (2,1)	$\frac{1}{18}$
Sum = 4	(1,3),(3,1),(2,2)	$\frac{1}{12}$

Sum = 5	(2,3),(3,2),(1,4),(4,1)	$\frac{1}{9}$
Sum = 6	(1,5),(5,1),(2,4),(4,2),(3,3)	$\frac{5}{36}$
Sum = 7	(1,6),(6,1),(2,5),(5,2),(3,4),(4,3)	$\frac{1}{6}$
Sum = 8	(2,6),(6,2),(3,5),(5,3),(4,4)	$\frac{5}{36}$
Sum = 9	(3,6),(6,3),(4,5),(5,4)	$\frac{1}{9}$
Sum = 10	(4,6),(6,4),(5,5)	$\frac{1}{12}$
Sum = 11	(5,6),(6,5)	$\frac{1}{18}$
Sum = 12	(6,6)	$\frac{1}{36}$
Even Sum	Sum = (2,4,6,8,10)	$\frac{1}{2}$
Odd Sum	Sum = (1,3,5,7,9)	$\frac{1}{2}$
Prime	Sum = (2,3,5,7,11)	$\frac{5}{12}$
Composite	Sum = (4,6,8,9,10,12)	$\frac{7}{12}$

2. 5 Card Poker

5 Card Poker Hand (One 52 Card Deck)	Probability
Royal Flush	$\dfrac{1}{649,740}$
Straight Flush	$\dfrac{3}{216,580}$
Four of a Kind	$\dfrac{1}{4,165}$
Full House	$\dfrac{6}{4,165}$
Flush	$\dfrac{1,277}{649,740}$
Straight	$\dfrac{5}{1,274}$
Three of a Kind	$\dfrac{88}{4,165}$
Two Pair	$\dfrac{198}{4,165}$
Pair	$\dfrac{352}{833}$

3. Yahtzee

Yahtzee Roll	Probability
Yahtzee	$\dfrac{1}{1,296}$
Four of a kind	$\dfrac{25}{1,296}$
Full House	$\dfrac{25}{648}$
Large Straight	$\dfrac{5}{162}$
Small Straight	$\dfrac{10}{81}$

Three of a kind	$\dfrac{25}{162}$
Chance	$\dfrac{205}{324}$

4. Lotto

Given a Lotto Drawing with p picks from n total picks, our probability

is $P = \dfrac{p!}{\dfrac{n!}{(n-p)!}}$

5. Roulette

Roulette Bet	Probability	Expected Return on $1 bet
Red	$\dfrac{9}{19}$	$-0.05
Black	$\dfrac{9}{19}$	$-0.05
Odds	$\dfrac{9}{19}$	$-0.05
Evens	$\dfrac{9}{19}$	$-0.05
1-12	$\dfrac{6}{19}$	$-0.37
13-24	$\dfrac{6}{19}$	$-0.37
25-36	$\dfrac{6}{19}$	$-0.37
One Number	$\dfrac{1}{38}$	$-0.95
Two Numbers	$\dfrac{1}{19}$	$-0.89
Four Numbers	$\dfrac{2}{19}$	$-0.79

CHAPTER 52

RESOURCES
AND PRODUCTS

I can be found using the following channels:

- Email: don@mathcelebrity.com
- Twitter: MathCelebrity

I sell the following products and services at MathCelebrity

- Introductory Programming Course
- Math Based Algorithms Course
- Pattern Recognition Course
- Free Traffic Secrets course. This contains the content mentioned in the "How to get 200,000 monthly visitors chapter" as well as many more tips I've learned over the years
- Personal Homework Coach
- Consulting Services

You can find these at http://www.mathcelebrity.com/products.php

Our friends and affiliates can be found at MathCelebrity Friends and Affiliates

http://www.mathcelebrity.com/charity.php

onesecondmath.com

www.1secondmath.com

CHAPTER 53

GLOSSARY

HTML API Positive Feedback Loop PHP

Glossary

API   Application Programming Interface (API) is a particular set of rules and specifications a software program follows to access and make use of the services and resources provided by another particular software program implementing that API.

HTML   Hyper Text Markup Language: A web programming language

PHP   Hyper Text Pre Processor: A server side programming language

Positive Feedback Loop   A positive feedback loop is where you do something, it works, and you do more of it which in turn feeds more momentum to your behavior

www.ingramcontent.com/pod-product-compliance
Lightning Source LLC
Chambersburg PA
CBHW071534200326
41519CB00021BB/6480